Arthur Erwin Brawn

Guide to the garden of the Zoological Society of Philadelphia

According to the present Arrangement

Arthur Erwin Brawn

Guide to the garden of the Zoological Society of Philadelphia
According to the present Arrangement

ISBN/EAN: 9783741184154

Manufactured in Europe, USA, Canada, Australia, Japa

Cover: Foto ©berggeist007 / pixelio.de

Manufactured and distributed by brebook publishing software
(www.brebook.com)

Arthur Erwin Brawn

Guide to the garden of the Zoological Society of Philadelphia

OF THE

OOLOGICAL SOCIE'

OF PHILADELPHIA,

(FAIRMOUNT PARK,)

ACCORDING TO THE PRESENT ARRANGEMENT

BY ARTHUR ERWIN BROWN,

General Superintendent.

THE ZOOLOGICAL SOCIETY

OF PHILADELPHIA

Was incorporated in 1859. The Garden was first opened to the public July 1st, 1874.

It has four different classes of members, viz.:—

1. ANNUAL.—These pay five dollars upon their election, and five dollars for each year thereafter.

2. LIFE MEMBERS.—These pay fifty dollars upon election, in lieu of all future dues. Annual Members may at any time become Life Members upon the payment of forty-five dollars additional.

3. HONORARY MEMBERS are those who, in consequence of liberality to the Society, or who hold a distinguished position in science, are elected by the Board.

4. CORRESPONDING MEMBERS are those who are interested in the Society, living out of the city, and are of service to it abroad.

Persons who wish to become members will please communicate with the Secretary of the Society.

All members are admitted to the Garden during the time it is open to the public, which is from 9 A. M. until sunset, every day in the week, including Sunday.

PLAN

OF THE

PHILADELPHIA ZOOLOGICAL GARDEN.

1878.

18

CARNIVORA HOUSE

A

C

D

Girard   Avenue

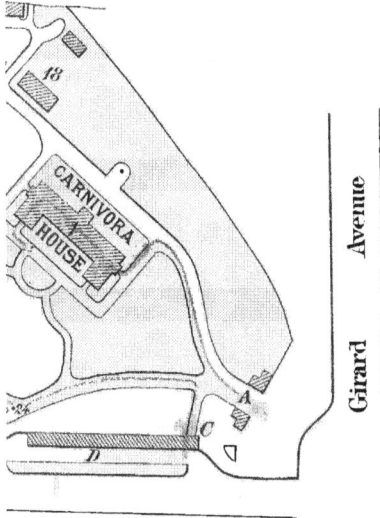

19. GOATS, SHEEP, OSTRICH, &c.
20. SNAKE HOUSE.
21. AVIARY.
22. PRAIRIE DOGS.
23. PHOTOGRAPH STAND.
24. UNIVERSAL SUN DIAL.
25. SODA FOUNTAINS.
26. RABBIT WARREN.
27. POLAR BEAR PIT.
28. SEAL TANK.

PLAN
OF THE
PHILADELPHIA ZOOLOGICAL GARDEN.
1878.

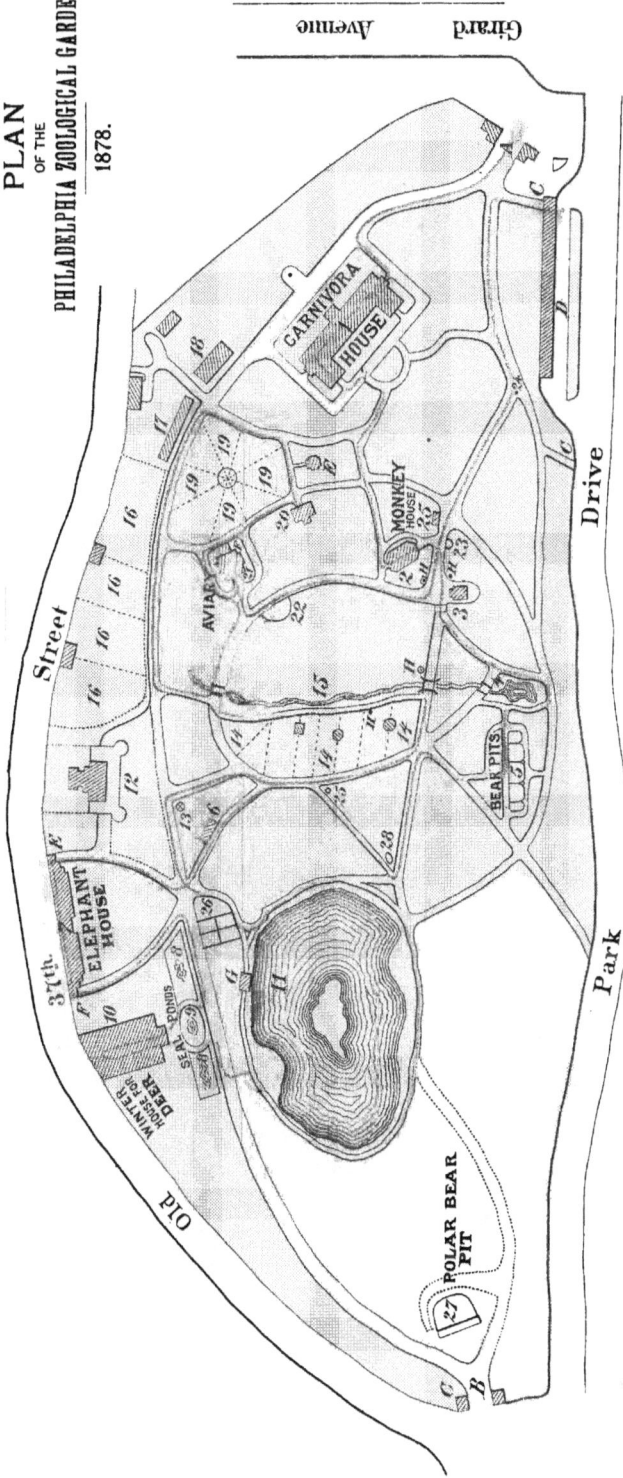

A. NORTHERN ENTRANCE.
B. SOUTHERN ENTRANCE.
C. EXIT GATES.
D. CARRIAGE SHEDS.
E. RETIRING ROOMS FOR MEN.
F. RETIRING ROOMS FOR WOMEN.
G. SKATING HOUSE.
H. DRINKING FOUNTAINS.

1. CARNIVORA HOUSE.
2. MONKEY HOUSE.
3. PENN MANSION, OR SOLITUDE.
4. BEAVER POND.
5. BEAR PITS.
6. EAGLE, HAWK AND OWL-CAGE.
7. ELEPHANT HOUSE.
8. } SEAL PONDS.
9. }

10. WINTER HOUSE FOR DEER, &c.
11. LAKE FOR WATER FOWL.
12. RESTAURANT.
13. MUSIC STAND.
14. DEER PADDOCKS.
15. STREAM FOR WATER FOWL.
16. BUFFALO, CAMELS AND ELK.
17. WOLVES AND FOXES.
18. CONSERVATORY.

19. GOATS, SHEEP, OSTRICH, &c.
20. SNAKE HOUSE.
21. AVIARY.
22. PRAIRIE DOGS.
23. PHOTOGRAPH STAND.
24. UNIVERSAL SUN DIAL.
25. SODA FOUNTAINS.
26. RABBIT WARREN.
27. POLAR BEAR PIT.
28. SEAL TANK.

OF THE

ZOOLOGICAL SOCIETY

OF PHILADELPHIA,

(FAIRMOUNT PARK,)

ACCORDING TO THE PRESENT ARRANGEMENT.

BY ARTHUR ERWIN BROWN,

General Superintendent.

PRICE, FIFTEEN CENTS.

To a large class of visitors, who desire to find in a zoological collection means of instruction as well as of amusement, à brief statement of the meaning and the relative value of the groups into which the animal kingdom is divided by naturalists will not be without interest. In order to arrive at a correct understanding of these, it is necessary to look at the animal world, not as a mere mass of living forms bearing hap-hazard resemblances to each other, but as great family groups of beings, formed, more or less, on the same plan, varying, it is true, to a vast extent in the manner of its manifestation, but all the forms of which are capable of being arranged around great centres, each of which presents a somewhat different combination of structural peculiarities.

All the living forms known to science were first systematically arranged by Linnæus, and though the researches of later naturalists have at times classified them on other bases and in different ways, there is now a tendency to return, to a certain degree at least, towards the system of the great Swedish naturalist.

The primary divisions now generally accepted are as follows :—

The *Vertebrata*—possessed of a backbone, as mammals, birds, reptiles, batrachians, and fishes.

The *Mollusca*—soft-bodied, as oysters, cuttle-fishes, and snails.

The *Articulata*—formed of rings, as worms, centipedes, insects, &c.

The *Cœlenterata*—as sea anemones and jelly-fishes.

The *Echinodermata*—as star-fishes and sea cucumbers.

The *Protozoa*—as sponges and infusoria—the lowest forms of animal life, many of them microscopic and bordering closely on the vegetable world.

It is with the first division only that the collection in the Garden has to deal. The Vertebrates—animals possessing a skeleton of bone or cartilage, enclosing cavities in which the soft parts of their organization are contained and protected from injury—are arranged in five *classes*, according to the nearness with which they approach to one of the five great types of structure which have been found to exist among them:—

I. *Mammalia*—animals which suckle their young.
II. *Aves*, or birds.
III. *Batrachia*—as frogs, toads, and salamanders.
IV. *Reptilia*—as turtles, lizards, and serpents.
V. *Pisces*, or fishes.

These *classes* are again broken up into *orders*, each possessing an association of structural characters which is common to all the individuals included in it, and in which they differ from all other individuals in their *class*. These *orders* have been differently constituted and arranged, according as different points have been made use of for their determination.

They are again divided into smaller groups called *families*, which, possessing the characteristic mark of their *order*, yet depart in some minor consideration from its type—or, in other words, from that form which has been taken to show most clearly the peculiarity of the order.

Families are again broken up into *genera*, which bear to them much the same relation as that which they, in turn, bear to orders. Thus—to illustrate with a familiar example—the lion, tiger, panther, &c. are all cats and belong to one genus—*Felis;* they are classified as follows :—

Division *Vertebrata*—because they have a backbone or vertebral column.

Class *Mammalia*—because they have organs peculiar to those animals which suckle their yonng.

Order *Carnivora*—because their plan of structure is that possessed by animals which live on flesh.

Family *Felidæ*—because, in addition to the above, they possess a common arrangement of teeth, claws, and other structural points, which none of the other carnivora share.

Genus *Felis*—because certain minor modifications are unlike those existing in a few more individuals, which so far have agreed with them, but which now becomes another genus of the same family.

But to go a step farther—the lion, tiger, and panther, though so far they have been precisely similar, are yet recognizable—there are still smaller points of difference ; they are, therefore, said to be different *species*, and a second name is added to the scientific designation of their genus ; thus the Lion is *Felis leo*, the Tiger is *Felis tigris*, and the Panther is *Felis concolor*. The value of species has been admirably expressed by Professor Huxley:—"Thus horses form a species, because the group of animals to which that name is applied is distinguished from all others in the world by the following constantly associated characters :—They have—1. A vertebral column ; 2. Mammæ ; 3. A placental embryo ; 4. Four legs ; 5. A single well-developed toe on a foot provided with a hoof ; 6. A bushy tail ; and 7. Callosities on the inner side of both the fore and hind legs. The asses, again, form a distinct species, because, with the same characters, as far as the fifth in the above list, all asses have tufted tails, and have callosities only on the inner side of the forelegs. If animals were discovered having all the general characters of the horse, but sometimes with callosities only on the forelegs, and more or less tufted tails, or animals having the general characters of the ass, but with more or less bushy tails, and sometimes with callosities on both pairs of legs, besides being intermediate in other respects, the two species would have to be merged into one. They could no longer be regarded as morphologically distinct species, for they would not be distinctly definable one from the other."—*Westminster Review, April, 1860.*

GUIDE TO THE GARDEN

OF THE

ZOOLOGICAL SOCIETY

OF PHILADELPHIA.

THE visitor taking the route laid down on the accompanying plan of the Garden, is supposed to enter at the Girard avenue gate ; those entering at the other end of the Garden can, however, pursue the same course by taking the path to the right around the lake (No. 11), and following the route until they come to the Sun-Dial (No. 24), when they should turn to the left and enter the Carnivora House, after which the route can be followed until they are brought back to the point from which they started. As many of the animals are shifted from one place to another at different seasons, they will not always be found at the location designated by the Guide ; each cage, however, bears the name of the animal which it contains, and its description can readily be found by a reference to the index at the close of the book.

No. 1.—THE CARNIVORA HOUSE.

THE step appears to be a long one from the domestic tabby, which is accustomed to lie purring before the fire, to the majestic Lion, which the visitor sees with much pleasure is here separated from him by a solid framework of iron ; yet there is no difference between them so far as the essential points of their structure are concerned, and none even in their habits, excepting such as are caused by the different circumstances under which they live. The one preys on deer and antelope, while the other lives on rats and mice; but

CARNIVORA BUILDING

they hunt for them and catch them in one and the same manner. They are the two extremes of the *Felidæ*, or cat family, regarded in point of size.

The true cats—composing the genus *Felis*—are externally distinguished from the other members of the family, or the lynxes, by their more slender form and by a much longer tail than is ever possessed by any of the latter genus.

They are distributed throughout America, Asia, and Africa; one species alone being found in Europe.

THE LION (*Felis leo*) ranges all through Africa, from the Cape of Good Hope to the Mediterranean and through a great part of Southern Asia, but there, at least, in rapidly lessening numbers.

In ancient times, we are told by historians, they were known in Greece, but civilization has long since driven them out of Europe.

THE LION AND LIONESS.

As with all animals whose wide range of distribution exposes them to many different climates and kinds of food, under the action of which, with other conditions, they are apt to vary more or less, the Lion of different parts differs much in appearance, chiefly in color and thickness of mane. These differences were formerly held to constitute distinct species, but as the variations are indefinite and do not involve the slightest change of structure, they are now looked on as being of no specific importance.

The large male, now in the Garden, is a fine specimen of the African Lion in the prime of life, with all the characters of his race fully developed. One of the two Lionesses, also in the collection, is the mother of the pair of well-grown cubs, "Simon Kenton" and "Daniel Boone," born in Louisville, Kentucky, and now also belonging to the Society; they will not reach their full development until their eighth year.

In the south-east wing of the building is another member of the family, several years younger.

A·L·S·

THE TIGER.

THE TIGER (*Felis tigris*), among the Carnivora, is the sole rival of the lion, in strength and ferocity. Its range is much more restricted, as it is never found outside of Asia, where its principal home is in Hindostan and the adjacent islands, though it is sparingly found towards Siberia on the north and China on the north-east. It is very common in the marshy, wooded tract known as the Soonderbund, formed by the extensive delta of the Ganges and Brahmapootra rivers.

There are two fine males in the collection, to the largest of which a somewhat tragic interest attaches, as shortly after he came into the possession of the Society, in May, 1876, he inflicted injuries upon his mate, from the effects of which she shortly died.

The male and female Tiger are similar in appearance, and have been beautifully adapted by nature for the purpose of stealing unobserved upon their prey; the tawny yellow of their skin, striped with vertical bars of black, blending so perfectly with the jungle of canes and bamboos, among which they live, that it is almost impossible to detect their presence until revealed by motion, when it is usually too late for the startled victim to escape.

The Leopard has much the same distribution as the lion and varies almost to the same extent. This building contains a pair of the COMMON LEOPARD (*Felis pardus*) of Africa and Asia. The male of this pair distinguished his arrival at Philadelphia from Hamburg by breaking out of his cage and taking possession of the hold of the canal-boat in which he had been brought from New York. For three days he maintained an obstinate defense, but hunger finally got the better of him and he fell a victim to the wiles of his keeper. The barge being named "The Chesapeake," the animal was at once called "Commodore Lawrence," in honor of his gallant naval predecessor.

THE JAVAN LEOPARD (*Felis pardus var. javanensis*), sometimes called the SPOTTED PANTHER, is also represented in the collection by a male and female. In the next cage to these is a fine pair of BLACK LEOPARDS (*Felis pardus var. melas*). This a rare variety in color of the common Leopard, and is believed to exist only in Java.

THE COMMON LEOPARD.

THE JAGUAR (*Felis onça*) is the largest of the cats of the New World ; it inhabits the hottest parts of the continent from South America into Upper Mexico, sometimes even ranging into the United States,—the principal home of the species being in the dense forests which stretch away from both banks of the Amazon.   In appearance there is much similarity to the leopard,—the Jaguar having a shorter tail and a more broken appearance of the spots covering the skin.

Many instances are given by South American travelers of the strength and ferocity of the Jaguar; D'Azara, in particular, relates how he once saw one drag off the body of a horse to a considerable distance and then swim with it across a wide and deep river.

THE AMERICAN PANTHER (*Felis concolor*) is generally distributed through North and South America from Canada nearly to Cape Horn, though in the most settled portions of the former, civilization has generally driven it to the secluded parts of the mountains of the North and East, and the cane brakes of the South.

They are found of several shades, from silvery gray to reddish brown, and are all of one species, though known by the different names of Panther, Puma, Couguar, and Mountain Lion. The latter name was given to them by the early colonists of the country, probably for the reason that the Panther, having no mane and approaching in color to the lion, was taken to be a female of that species, which is also devoid of a mane.

THE OCELOT (*Felis pardalis*) is a native of Mexico, Central and South America, and occasionally those parts of the United States bordering on Mexico. It is a beautiful animal, not much larger than the domestic cat. Like all of the cat family, with the exception of the lion and tiger, it climbs trees with great agility and lies in wait among the branches for its prey.

THE SPOTTED HYÆNA.

THE SPOTTED HYÆNA (*Hyæna maculata*) and the STRIPED HYÆNA (*Hyæna striata*) are members of the Family *Hyænidæ*, consisting of themselves and the BROWN HYÆNA, of which the Society does not, as yet, possess a specimen. They are readily distinguished from the dogs, which they somewhat resemble, by the excess of length in the fore over the hind limbs. Their molar teeth are unusually strong and the jaw muscles are very powerful, thus enabling them to crush with ease large bones, which they devour.

The Striped Hyæna inhabits both Africa and India, and presents a marked appearance by reason of a mane or crest of hair, running the length of the spine, and which it has the power to raise at will, probably for the purpose of increasing its bulk and giving to itself a more frightful appearance, thereby deterring from an attack those enemies which would otherwise overcome and destroy it. It is more cowardly in disposition and solitary in habit than the Spotted Hyæna, which is confined to the southern part of Africa.

A great deal is currently believed of the Hyæna which is without doubt much exaggerated; for instance, its reputation of being a persistent and incorrigible ghoul, which has passed so generally into belief that skulls and tombstones are usually introduced as background in portraits of the unfortunate animal—the truth being that the Hyæna seems to be closely on a par with the dogs and wolves in the matter of diet, preferring his bones fresh and eating carrion only when it is much more convenient to get at.

The *Viverridæ* is a large family of carnivorous mammals of small size, all resembling, more or less, the Civets, in appearance and habits; they are of active and graceful movements, many of them living much among trees; all feeding upon smaller quadrupeds, birds, eggs, and reptiles. They chiefly inhabit Africa and Southern Asia, one species each being found in Europe and America.

Among them are the INDIAN CIVET (*Viverricula indica*), the PALM CAT (*Paradoxurus musanga*), the COMMON PARADOXURE (*Paradoxurus typus*), the GRAY ICHNEUMON (*Herpestes griseus*) from India and the surrounding islands, and the beautiful CIVET CAT or RING-TAILED BASSARIS (*Bassaris astuta*) of Texas and Mexico. It is probable that the latter animal possesses affinities which ally it more closely with the

coatis and raccoons than with the *Viverridæ*, among which it
has heretofore been classed.

THE COATI (*Nasua nasica*) will be readily recognized by
its long, pointed snout. There are two varieties, the Red and
the Brown Coati, though they are probably of one species—
native to Mexico, Central and Upper South America. It is
worthy of note that the first wild Coati ever found within the
limits of the United States was captured in 1877, near Fort
Brown, Texas, by Dr. Merrill, U. S. A. This species is allied
by many points of structure to the bears and raccoons, and has
been placed in the same family as the latter.

The BORNEAN SUN BEAR (*Helarctos euryspilus*) and the
HIMALAYAN BEAR (*Ursus tibetanus*) belong to a group of the
Ursidæ known as sun bears, from their favorite habit of bask-
ing in the sun. They are in the same cage, but may be readily recognized by the mane and larger size of the Himalayan
Bear and also by the V-shaped spot on the breast, which is
white in this and orange in the Bornean Bear.

The *Rodentia* is a very large order, characterized by the absence of canine teeth and the development of the incisors to
so great a degree that they resemble chisels and are used by
the animal for the purpose of cutting wood and other hard
substances, from which is derived their name—*Rodentia* or
gnawers. Representatives of this order are found all over the
world, North America having a large proportion of the whole
number of species. During the winter a number of small
cages are set in the wings of this building, which in warm
weather are scattered around the grounds, the occupants of
these belong mostly to this order ; among them are generally
the GOLDEN AGOUTI (*Dasyprocta aguti*); the OLIVE AGOUTI
or ACOUCHY (*Dasyprocta acouchy*) from South America and
the West Indies; the PACA or SPOTTED CAVY (*Cælogenys paca*),
and FOURNIER'S CAPROMYS (*Capromys pilorides*).

The AFRICAN PORCUPINE (*Hystrix cristata*), the JAVAN
PORCUPINE (*Hystrix javanica*), the WHITE-HAIRED PORCUPINE
(*Erethizon dorsatus*), and the YELLOW-HAIRED PORCUPINE
(*Erethizon dorsatus var. epixanthus*)—the two last from North
America—are all quiet, retiring Rodents, living on roots and
vegetables or the bark of trees. The spines which take, in
part, the place of hair in the Porcupine, are loosely rooted in
the skin and readily come off in the mouths of such animals

as may attack them, thus forming a terrible means of defense to the animal. The ease with which these spines are detached has, without doubt, given rise to the fable that the Porcupine was able to shoot forth its quills, like arrows, against its foes.

THE FLYING SQUIRREL (*Pteromys volucella*) is a pretty little Rodent found throughout the United States, east of the Missouri river. Its aerial progression is merely a leap, prolonged by means of a fold of skin stretching between the fore and hind limbs on each side, which expands and bears the animal up for a short distance, after the manner of a kite.

THE VULPINE PHALANGER (*Phalangista vulpina*) and the YELLOW-BELLIED PHALANGER (*Belideus flaviventer*) are small, vegetable-eating marsupials (see page 41) from Australia ; they live almost entirely among the trees and are strictly nocturnal, being found in the day time with heads bent down and noses stowed away between their forefeet.

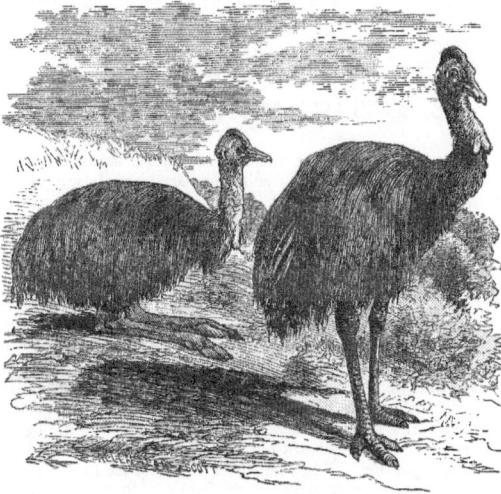

THE COMMON CASSOWARY.

The Cassowary is also kept in this building during the winter, in summer finding quarters in a cage on the walk toward the Monkey House. There are some half dozen species of the genus *Casuarius*, mainly differing in the shape of the helmet on the head and the number and arrangement of the wattles hanging from the neck; all are natives of the islands of the Malayan Archipelago. They belong to the order of struthious birds, with the ostrich, rhea, and apteryx, all of which are characterized by great development of the lower limbs at the expense of their powers of flight.

THE COMMON CASSOWARY (*Casuarius galeatus*) is from the island of Ceram, in the Indian Ocean. The feathers of this bird are of a peculiarly filamentous or hair-like character, entirely wanting in the webs which spring from the sides of the shaft in ordinary feathers. It is a bird of great power and endurance, rivaling even the ostrich in those qualities as well as in those famous powers of digestion which are so notorious in the latter bird.

No. 2.—THE MONKEY HOUSE.

THE present house for Monkeys has been found too small and badly ventilated for the proper accommodation of the animals, and the Society has in contemplation the erection of a new and suitable building when circumstances will warrant its completion.

The Monkeys of the Old World, or of Africa, Asia, and the Malayan Islands, have been arranged by naturalists in one great group called *Catarrhini*, while those of the New World constitute another group known as *Platyrrhini*. They are very well marked in zoological characters, the most constant of which is that from which they derive their name. In the *Catarrhini* the septum, or cartilage dividing the nose, is narrower at the bottom than at the top, so that the nostrils converge towards the bottom, while in the *Platyrrhini* the cartilage is of the same breadth throughout and the nostrils are therefore parallel. The dentition of the first group is the same as that of man, being eight incisors, four canines, and twenty molars.

The Monkeys are classed in the same order—*Primates*—as
man, the correspondences in mere structure being very close,
amounting, in some of the higher groups, to modifications
only of form.   All the man-like or Anthropoid Apes—the
Gorilla and Chimpanzee of Africa, and the Orang and Gib-
bon of India—belong to the first division.   These Apes can
be captured only when young, and as they are difficult to ac-
climate, they are by no means common in menageries.   The
Society possesses a pair of CHIMPANZEES (*Troglodytes niger*),
which arrived in April from the Gaboon river, in Africa.

The species inhabits the west coast of Africa near the Equator; its exact range, north, south, and inland, is not satisfactorily determined, but it is probably confined to a limited region in company with its larger relative, the Gorilla.

Divesting the Chimpanzee of the many doubtful, if not fabulous qualities with which it has been endowed by imaginative travelers, it remains a huge Ape, attaining in the adult male a height of four and a half feet; devoid of a tail; possessed of a very considerable degree of intelligence, and having the ability to walk upright, supporting itself by occasionally touching its knuckles to the ground or some upright means of support.

They live together in small bands of half a dozen and build platforms among the branches, out of boughs and leaves, on which they sleep; their diet is chiefly frugivorus, and they are exceedingly mild in disposition, readily becoming friendly and seeking the society of man when placed in confinement.

These Apes are looked on by the natives of their country as being degenerate members of their own tribe. The native name, " Engeco, " means " hold your tongue," and evidently had its origin in the common belief that they refuse to speak purely from laziness, and in the fear that if their possession of the faculty should be discovered, they would be set to work with the more bipedal inhabitants of the same region. The pair in the Garden are believed to be about four years old,—their development is slow, as it is not supposed that they reach maturity until about fifteen years of age.

Among the Monkeys of the Old World which are usually to be found in the collection, is the ENTELLUS or SACRED MONKEY (*Semnopithecus entellus*) of India. This Monkey is held in high respect by the human natives of its country, who call it HANUMAN, after one of their deities, and allow it the privilege of stealing, unmolested, pretty much anything to which it takes a fancy—a privilege which it soon learns to avail itself of on every occasion. Its life is held sacred, and it is a dangerous thing for a foreigner to incur the displeasure of the people by killing one.

THE VERVET MONKEY (*Cercopithecus lalandii*), the GREEN MONKEY (*Cercopithecus callitrichus*), the PATAS or RED MONKEY (*Cercopithecus ruber*), the LESSER WHITE-NOSED MONKEY

GROUP OF MONKEYS.

(*Cercopithecus petaurista*), and the MANGABEY (*Cercocebus fuliginosus*) are all natives of South and West Africa.

The Macaques form a large genus of Monkeys, some of them of large and powerful build, and, for the most part, of savage and treacherous dispositions. They are all natives of Asia and the adjacent islands. Among them are the COMMON MACAQUE (*Macacus cynomolgus*), the PIG-TAILED MACAQUE (*Macacus nemestrinus*), the RHESUS MONKEY (*Macacus erythræus*), the TOQUE (*Macacus pileatus*), and the BONNET MACAQUE (*Macacus radiatus*).

The genus *Cynocephalus*, or Dog-headed Monkeys, form the group known as Baboons, among which are some of the largest and most fierce of the order. THE CHACMA (*Cynocephalus porcarius*), the GUINEA BABOON (*Cynocephalus sphinx*), the MANDRILL (*Cynocephalus mormon*), the DRILL (*Cynocephalus leucophæus*), and the ANUBIS BABOON (*Cynocephalus anubis*), are all natives of Africa. They can all be recognized by their long, dog-like noses, in some cases projecting beyond the lips.

Although these Monkeys are coarse and brutal in their behavior towards man, they are capable of a high degree of attachment among themselves.

A remarkable instance of this is given by Brehm, an African traveler of undoubted veracity, who once saw a troop of Baboons crossing a valley,—while so doing they were attacked by his dogs, and fled up the hills, leaving behind one young one, which, unable to run away, had climbed a rock in the middle of the valley. Those on the hillside deliberated for a time, and finally a large male returned to the spot, drove off the dogs, picked up the young one, and retreated with it in safety.

The American Monkeys differ in many respects from the preceding group; in dentition, which in the *Cebidæ*, including all but the Marmosets, has one pre-molar tooth added on each side of the jaw; in the absence of a thumb in almost all the members of one large genus (*Ateles*); in the entire absence of the callosities on the haunches, which are so conspicuous in most of the *Catarrhini*; and in the presence of a highly prehensile tail in the individuals of all the leading genera. None of them attain the size of the largest of the first group, and they are generally more tractable in disposition.

THE ATELES or SPIDER MONKEYS are characterized by the absence of a thumb, although in several species it is present in a rudimentary condition; they have a prehensile tail, lined on the tip with a very sensitive skin, which answers the purpose of a hand in suspending themselves from the branches of the trees among which they altogether live. They are very delicate and do not long withstand the severities of our climate.

THE BLACK SPIDER MONKEY (*Ateles ater*), the MARIMONDA (*Ateles belzebuth*), and the BLACK-HANDED SPIDER MONKEY (*Ateles melanochir*), are usually to be seen here. A fine specimen of the latter from Central America, "Jerry," passes much of his time during the summer at large among the trees in the neighborhood of the Monkey House, and creates much amusement by his antics; he seems never disposed to take advantage of his freedom to run away, and manifests a most devoted attachment to his keeper.

THE BROWN CAPUCIN (*Cebus fatuellus*), the WEEPER CA-
PUCIN (*Cebus capucinus*), and the WHITE-THROATED CAPUCIN
(*Cebus hypoleucus*), are all small Monkeys of the kind usually
trained for circus performances and organ-grinders. These,
with the SQUIRREL MONKEY (*Saimaris sciurea*), all belong to
the family *Cebidæ*.

The COMMON MARMOSET (*Hapale jacchus*), the BLACK-
EARED MARMOSET (*Hapale pencillata*), and the PINCHE (*Mi-
das œdipus*), are small and beautiful Monkeys from the hottest
parts of tropical America.

The Monkeys of the New World range from about fifteen
degrees north to forty degrees south latitude; the most northern point which they reach in the eastern hemisphere being
in Tunis, about thirty-eight degrees north latitude.

Leaving the Monkey House, the visitor passes the old mansion, "Solitude," erected in 1785 by John Penn, a descend-
ant of the founder of the Commonwealth, and now occupied
by the offices of the Society—and descending a flight of steps
turns to the left by

No. 4.—THE BEAVER POND.

THE identity of the Beaver of North America with that of
Europe has been for many years a subject of discussion
among naturalists, and by a large number their specific dis-
tinction is considered as assured. Recent researches, however, embracing the comparison of a large number of skulls,
place beyond dispute the fact that the cranial characters,
which were taken to warrant the separation of the two forms,
are subject to so great an amount of variation in different individuals that they cannot be considered as binding. It
seems proper, therefore, that the American form should be
looked on merely as a variety of that from the Old World.

The AMERICAN BEAVER (*Castor fiber var. canadensis*) has
been so valuable in times past to commerce, that a considerable amount of interest has been felt in the organization of
their villages, which are said to manifest a degree of system
almost beyond anything else to be found among the lower
animals.

The Beaver Pond at the Garden affords, perhaps, as good an opportunity as is possible, in confinement, of watching the habits of these animals,—the rough, dome-shaped structure of mud and sticks on the island being the far-famed Beaver hut, built by these animals themselves out of the natural materials. In a state of nature these huts are generally built on a small stream where the Beaver have constructed a dam, deepening the water sufficiently to keep it from freezing to the bottom, so that they can get out under the ice during the winter. Most of their work is done during the night, but towards the hour in the afternoon when they are accustomed to be fed they may often be seen swimming about the pond and disappearing with a flap of the tail as they get within diving distance of the entrance to the hut.

They have done very well in their present quarters, and have bred there during the past summer. Their food is purely vegetable, consisting mostly of the bark of trees and the roots of aquatic plants. They formerly ranged over the whole of North America, but have long since been exterminated in the Southern, and in great part in the Middle and Eastern States.

No. 5.—THE BEAR-PITS.

But three well-marked species of Bears are believed to exist in North America—the GRIZZLY BEAR (*Ursus horribilis*); the BLACK BEAR (*Ursus americanus*), with its constant variety (*cinnamomeus*); and the POLAR BEAR (*Ursus maritimus*)—though the variations of both the Grizzly and the Common Bear have given rise from time to time to supposed new species. The Grizzly ranges from the Yellowstone Valley and the Upper Missouri to California, and south-west into Mexico,—those from the western slope of the Rocky Mountains being somewhat darker in color and reaching a larger size than those to the eastward.

The Black Bear is found almost all over the country, together with what is called the Brown Bear, which is merely an ill-defined variety of the former species,—the black of the hair, in some individuals, turning to brown shortly after the yearly change of coat. The true Brown Bear is the *Ursus*

arctos of Northern Europe and Asia—an animal which presents many points of likeness to our Grizzly. The Cinnamon Bear is mainly confined to the north-western parts of the United States.

The *Ursidæ*, or Bears, all walk on the sole of the foot, or are what is termed *plantigrade*, and with the exception of the Grizzly, climb trees with great facility. The diet is much mixed, being indiscriminately animal or vegetable.

They are distributed throughout the Northern Hemisphere—but one true Bear having yet been found south of the equator.

The COMMON OPOSSUM (*Didelphys virginiana*) ranges east of the Missouri river, from about the latitude of lower Massachusetts to the Gulf of Mexico, and with its congener (*Didelphys californica*), which replaces it west of those limits, represents in North America the order of Marsupials. (See page 41.) The Opossums belong properly to the carnivorous branch of the order, although their diet is very varied, consisting of small birds, mammals, reptiles, and eggs, as well as of fruits, buds, and grain.

They live generally in the hollow of a tree, when the female produces as many as fifteen young at a time, breeding several times in the course of a year. The characteristic pouch of the order is well developed in the members of this group.

They have a very prehensile tail, and are also distinguished by the peculiarity of their dentition, which consists of ten incisors, two canines, and fourteen molars in the upper, and the same, with two incisors less, in the lower jaw, or fifty teeth in all.

A number of Opossums are found in South America, more or less resembling this species.

THE CONDOR (*Sarcorhamphus gryphus*) is the largest of the Vultures, rivaling and even exceeding in size the Bearded Vulture or Lammergeyer of the Alps.

They do not build nests, but commonly live in pairs on the bare rock, high up among the lofty peaks of the Andes, from which they soar to a height almost beyond the range of human vision, plunging down only when their keen sight discovers the carcass of some dead animal on the plains below. They live mostly on carrion, but when pressed by hunger, it is said that several of them will sometimes band together to attack a young calf or a disabled animal out of the herd, and with blows of their powerful beaks and claws destroy it. The sexes may be distinguished by a short ruff of soft feathers, which invests the neck of the adult male.

THE COMMON RACCOON (*Procyon lotor*). A number of these amusing little animals are in a cage nearly opposite the condor. They resemble in diet, and in many points of structure, the bears, and have been placed by some naturalists as a sub-family of the group. They are generally classed, however, as a separate family, *Procyonidæ*, of carnivores. Their range is almost universal through the United States, from the latitude of Massachusetts southward. They are subject to considerable variation in color—albinos being not uncommon. One specimen in the collection, from Alabama, is of an orange-yellow, shading into a deeper hue on those parts where the animal is normally black. They are easily tamed, and make playful pets. One which became very tame in the Garden was noticeable for the dexterity with which it made use of its paws—its first act on mounting into any person's lap being to explore all his pockets, bringing to light and carefully examining everything which they contained.

No. 28.—THE SEAL TANK.

THE Seals are a large family of carnivorous mammals, living mainly in the water, but at stated periods during the year leaving their natural element and remaining for several months above the water line. The Society has exhibited several species of these interesting animals.

THE COMMON SEAL (*Phoca vitulina*) is found in all the seas encircling Northern Europe, Asia, and America, rarely being found on our coast below Maine. It may be taken as a fair type of the *Phocidæ*, or Earless Seals, of which it is about the smallest. Those in the collection are from Nova Scotia, and were brought to the Garden when only a few weeks old. Like all the Seals, they live on fish, which, in a state of nature, they catch for themselves with great address. (For other members of the Seal family, see page 37.)

Along the walk from the Seal Tank to the Eagle Aviary are a number of small cages containing birds and small mammals. THE MUSKRAT (*Fiber zibethicus*) of North America, which by its subterranean galleries causes so much damage to the banks of canals and other artificial bodies of water, will be found here. Its habits are somewhat similar to those of the beaver.

THE COMMON CROW (*Corvus americanus*) and the RAVEN (*Corvus carnivorus*) are the leading American members of the family *Corvidæ*, represented in England by the magpie.

The fine pair of mysterious-looking Ravens in one of the cages on this walk were the gift of William Wister, Esq., and formed the nucleus of the collection. As they are very long-lived, it is probable that they will see greater changes in their surroundings than they have yet done. They are found throughout North America, although east of the Mississippi they have become rare.

THE FISHER, OR PENNANT'S MARTEN (*Mustela pennantii*), affords an example of a very prevalent and deplorable confusion among the vernacular names of animals, which is the cause of most erroneous ideas with regard to the habits of many species. There is no evidence whatever that this animal catches fish, or that it is particularly fond of a fish diet ; yet its most common name conveys the impression that these are its most noticeable habits. It belongs to the large family *Mustelidæ*, comprising the otters, weasels, skunks, &c.—all of which are carnivores of small or medium size, mainly living on land, though several of them, as the mink and otter, are essentially aquatic in their habits.

------·------

No. 6.—THE EAGLE AVIARY.

This building is divided into three compartments, containing Owls, Eagles, Hawks, and Vultures,—a number of these being also scattered around the neighborhood in small cages. Among them are generally to be found the following :—

THE GREAT HORNED OWL (*Bubo virginianus*), North America.

THE SCREECH OWL (*Scops asio*), North America.

THE BARRED OWL (*Syrnium nebulosum*), North America.

THE SHORT-EARED OWL (*Brachyotus palustris*), North America, Europe, and Asia.

THE SNOWY OWL (*Nyctea nivea*), Arctic regions.
THE BARN OWL (*Strix flammea var. americana*), United States.
THE JAVAN FISH OWL (*Ketupa javanensis*), Java.
THE TURKEY BUZZARD (*Cathartes aura*), America.
THE BLACK VULTURE (*Cathartes atratus*), Southern United States.
THE CARACARA BUZZARD (*Polyborus tharus var. auduboni*), Southern United States.
THE RED-TAILED BUZZARD (*Buteo borealis*), North America.
THE RED-SHOULDERED BUZZARD (*Buteo lineatus*), North America.
THE ROUGH-LEGGED BUZZARD (*Archibuteo lagopus var. sancti-johannis*), North America.
THE SPARROW HAWK (*Falco sparverius*), North America.
LANIER'S FALCON (*Falco lanarius var. polyagrus*), Western United States.
THE PIGEON HAWK (*Accipiter fuscus*), North America.
COOPER'S HAWK (*Accipiter cooperi*), North America.
THE BRAHMINY KITE (*Haliastur intermedius*), India.
THE GOLDEN EAGLE (*Aquila chrysætos*), North America.
THE BALD EAGLE (*Haliætus leucocephalus*), North America.

All the above belong to the order *Raptores*, or birds of prey, consisting of the Owls or nocturnal birds of prey (*Strigidæ*); the Eagles, Hawks, &c., or diurnal birds of prey (*Falconidæ*); and the Vultures (*Cathartidæ*).

The Owls are a large family, mainly of nocturnal habits, their eyes being adapted in structure to see in the dark, and the soft, downy plumage with which they are generally clothed enabling them to steal with noiseless flight upon the small birds, mammals, and reptiles which form their food.

They range generally throughout the world, differing somewhat in habit, but mostly in size.

The American Barn Owl is one of the most curious of the group. It abounds in the Southern States, and is frequently found as far north as New Jersey.

The large White or Snowy Owl is common to the more northern parts of both hemispheres. It moves somewhat south in winter, rarely getting below the latitude of New York.

The coloration of the Owls is generally indistinct, owing to the downy nature of their plumage, and is subject to an infinite amount of variation.

A large number of American species belong to the family *Falconidæ*.

The Buzzards proper are a group of Hawks, generally of large size and rather heavy flight. The Eagles are closely associated with this division. The Golden Eagle and the Bald Eagle are occasionally seen in all parts of the country, though they are now somewhat rare along the Atlantic coast, and for a long distance into the interior. They are the only Eagles properly belonging to the North American fauna, although as the Bald Eagle does not receive its white head and tail until its third year, its different stages of plumage have given rise to several vernacular names by which it is known.

Dr. Elliott Coues sums up the distribution and character of this species in his "Key to North American Birds" after the following descriptive manner:—

"North America, common; piscivorus; a piratical parasite of the Osprey; otherwise notorious as the Emblem of the Republic." Certainly, on watching its filthy habits, its sneaking ways, and its thievish expression, one is not disposed to be proud of his "Bird of Freedom."

The Falcons, Harriers, and Kites are small Hawks of rapid and vigorous flight and daring dispositions. They are found all over the world. The Falcons are readily distinguished by the presence of a notch—called the tooth—in the upper mandible.

The specimen of the DUCK HAWK (*Falco communis*) of North America, in one of the small cages, flew on board of the steamship Pennsylvania when three days out from the port of Philadelphia, and was added to the collection by the kindness of Captain Harris.

The true Vultures are represented in North America by three species, of which the Society possess two—the Turkey Buzzard, being common to all North America, and the Black Vulture, which is mainly confined to the South Atlantic and Gulf States, where they perform the duties of useful scavengers in the streets of even the large cities.

The pair of these birds, now in the Garden, were presented by the Mayor of Charleston, South Carolina.

The Caracara Buzzard belongs to a small group of Hawks approaching somewhat in habit to the Vultures. They range from South and Central America into the southern parts of the United States.

No. 12.—THE RESTAURANT.

THE Restaurant is on the right of the prescribed route, after leaving the Eagle Aviary, and offers to the hungry visitor ample facilities for supplying his necessities.

No. 7.—THE ELEPHANT HOUSE.

THIS building was completed in 1875, and cost about $38,000. It is one hundred and ninety-five feet long and affords ample accommodation for many of the larger animals.

It is the intention of the Society, at some future day, to enclose the grass-plot at the rear of this building with a heavy fence, and turn it into paddocks for the use of the Elephants and the Rhinoceros.

A large proportion of the animals in this building belong to the order *Ungulata*, or hoofed animals, comprising all in which the nail grows around the ends of the extremities and envelops them in a horny sheath known as the hoof. Some of these have one or three toes developed, while another group has two or four toes equally complete, the others being rudimentary. For purposes of convenience, therefore, the existing ungulates have been classed into two sub-orders, the *Perissodactyla*, or odd-toed, as the horse, rhinoceros, and tapir, and the *Artiodactyla*, or even-toed, comprising all the remaining hoofed animals, as deer, oxen, swine, &c. They are all vegetable eaters, and are found in all but the Australian region.

The JAVAN SWINE (*Sus vittatus*) and the ÆTHIOPIAN WART HOG (*Phacochœrus æthiopicus*) belong to the family *Suidæ*, or Swine. The latter remarkable-looking animal from Africa has several fleshy protuberances on the face, looking almost like horns. It is believed, from the observations of Mr. A. D. Bartlett, at the London Zoological Garden, that these warts have been developed by reason of their serving to protect the eyes from the upward strokes of the tusks in the desperate battles which the males wage against each other.

The Peccaries are not true swine, but they do not depart widely enough to be entirely separated from the group. The

COLLARED PECCARY (*Dicotyles torquatus*), ranges from the South-western United States into South America, and the WHITE-LIPPED PECCARY (*Dicotyles labiatus*), somewhat more southern in distribution and confined to South America. They are savage little animals, and, as they herd together in considerable numbers, they are regarded as dangerous enemies by both man and beast. They are not difficult to domesticate when taken young, but the presence of a pair of dorsal glands, secreting an unpleasant fluid which is apt to taint the meat after death, greatly lessen their value to man.

THE INDIAN ELEPHANT.

Several species of Zebras and Quaggas are found in Africa, and also of wild Asses in South-western Asia. The most beautiful of all these, in pattern and shade of coloration, is BURCHELL'S ZEBRA (*Equus burchelli*), which ranges in large herds over the plains north of the Orange river, in Africa. It is a curious fact, that some horses, especially those of a dun color, have indications of the spinal stripe and those on the inside of the legs, which are common among the Zebras, and which resemblance is held to indicate the relationship of the horse of the present epoch to some such antecedent form. A mouse-dun colored·work-horse recently belonging to the Society had these stripes plainly marked.

The Zebras are domesticated and tamed to some extent by the Boers, or farmers of South Africa.

The enormous animals which form the family called *Proboscidea*, from the peculiar elongation of their nose into a proboscis or trunk, are found at the present time in the warmer parts of Asia, in the Islands of Borneo, Sumatra, and Ceylon, and also in Central and Southern Africa.

There are two species, differing very appreciably,—the INDIAN ELEPHANT (*Elephas indicus*) has a concave forehead, comparatively small ears, and has four nails developed on the hind feet, while the AFRICAN ELEPHANT (*Elephas africanus*) has a rounder forehead, much larger ears, and has three nails on the hind foot instead of four. The incisor teeth, or tusks, as they are called, grow to an enormous size, but are rarely possessed by the female Indian Elephant.

The two large ones in this building are both females about eighteen years old, one being from Africa and the other from India. The two small ones, "Dom Pedro" and "Empress," respectively four and five years old, are both Indian, and were placed in the garden in December, 1876.

The Elephant is in reality a much smaller animal than is commonly supposed, careful measurements of large numbers, in India, showing that they average about nine feet in height at the shoulder, and rarely exceed ten. The ordinary life of the Elephant is supposed to be about a hundred years, although in special cases they undoubtedly live much longer.

It is given as a fact, on the authority of Sir Emerson Tennent, that the British, after their capture of the Island of Ceylon, in 1799, had in their service an Elephant which was proved by the records to have served the Dutch

AFRICAN ELEPHANT.

during the whole term of their occupancy,—something like a hundred and forty years.

The Elephant lives wholly on vegetable diet.

The INDIAN RHINOCEROS (*Rhinoceros unicornis*). There are several species of Rhinoceros found in Africa and Asia, the distribution of the animal being almost identical with that of the elephant. Almost all the species, with the exception of this and the Javan Rhinoceros, have two horns, one immediately behind the other. In the specimen in the Garden, only the stump of the horn is visible, as "Pete," being of a restless disposition, is inclined to rub his head against the walls of his cage, and so wears off his horn as fast as it grows out.

The thick hide of the Rhinoceros renders him almost invulnerable to the attack of other animals, and his great strength, which gives him an activity not at all in keeping

THE INDIAN RHINOCEROS.

with his appearance, coupled with the possession of a sharp,
strong horn on the bridge of his nose, causes him to be much
respected by the inhabitants of the region in which he lives.

"Pete" is usually very quiet, but is subject to attacks of
rage, during which he will turn violently on his keepers. On
such occasions he is suffered to remain by himself.

Three fine specimens of the GIRAFFE (*Camelopardalis gi-
raffa*) are in the possession of the Society at present.

Their native country is the central and eastern part of
Africa, from about the tenth or fifteenth degree of north
latitude almost to the Cape; here they are found in small
herds, browsing on the branches of such trees as may be
within their reach. Their long legs unfit them for feeding on
the ground, as it is a work of much difficulty for the Giraffe,
by straddling its fore limbs widely apart, to get its nose down
to the level on which it stands. Their gait is very rapid for

THE GIRAFFE.

a short distance, but their powers of endurance, or "bottom," as horsemen term it, not being proportionate, they soon settle down to a lesser rate of speed. The weapons of the Giraffe are its hoofs, with which it kicks and strikes in every direction, dealing powerful and dangerous blows. It is a very timid animal, however, keeping a careful lookout from its tall head, erected like a watch-tower, high in the air, and when startled by the approach of an enemy at once seeks safety in flight.

It is a ruminant, closely related to the antelopes. The horns, or protuberances on the head, are not shed, as is common with many of the order to which it belongs, but are composed of solid bone, covered with skin like the remainder of the skull.

As with the elephant, the height of the Giraffe is enormously over-estimated in popular opinion ; the distance from the head, when fully erect, to the ground, probably not averaging over sixteen feet. They are very delicate in constitution, and in our climate require the most careful attention.

The SOUTH AMERICAN TAPIR (*Tapirus terrestris*), as before stated, belongs to the same division of Ungulates as the horse and rhinoceros, though in appearance it somewhat resembles the swine. The natives of the regions which it inhabits consider it to be very good eating. It is fond of the water, diving and swimming with great ease, and is rarely found far from the banks of some lake or stream. Their common resort is the dense thickets of undergrowth, where they lie concealed from danger. D'Azara says of them :—" It is also remarked that when the jaguar pounces upon them, they rush headlong through the thickest parts of the woods, until they force him to quit his hold, passing through narrow and intricate places. The Mborebi, indeed, never frequents a beaten road or pathway, but breaks and pushes through whatever it encounters with its head, which it always carries very low. It flies all danger, and anticipates it by means of its strong nocturnal vision and its acute sense of hearing."

There are several not very well defined species in South and Central America and one in South-eastern Asia. Of this species (*Tapirus malayanus*) the Society obtained a specimen in the fall of 1876, but the severe winter which followed its purchase proved fatal to it, renewing a previous inflammation of the lungs, from the effects of which it died. It was much larger than the South American form and had a grayish-white patch marked out like a saddle-cloth over the back and sides from the shoulders to the rump—from this it derives its popular name of SADDLE-BACKED TAPIR.

During the winter a number of Macaws will be found in this building, which, when the weather is sufficiently warm, are kept in a wire cage opposite the Eagle Aviary.

These superb birds form an important group of the order *Psittaci*, or Parrots, and live in large flocks in the forests of Central and South America, where their brilliant colors vie in intensity with the tropical vegetation which surrounds them.

There are a number of species, of which the best known are the RED AND BLUE MACAW (*Ara macao*), the RED AND YELLOW MACAW (*Ara chloroptera*), and the BLUE AND YELLOW MACAW (*Ara ararauna*).

—————•♦•—————

Nos. 8 and 9.—THE LARGE SEAL PONDS.

THE upper one of the three large ponds is at present unoccupied, it being the desire of the Society to fill it with a colony of otter, which, owing to the difficulty of procuring them uninjured, it has not yet been possible to obtain.

The central and lower ponds are tenanted by a number of GILLESPIE'S HAIR SEALS (*Zalophus gillespii*). This species is found in large numbers in the lower part of the North Pacific Ocean ; those which are in the Garden having been captured at the San Miguel Islands, off the coast of California, not far from Santa Barbara ; they are rarely seen as far up as San Francisco, and are found in the waters of the same latitude on the Asiatic side of the Pacific.

The differences between this species and the NORTHERN SEA LION (*Eumetopias stelleri*), numbers of which afford great amusement to visitors to Seal Rock, in the bay of San Francisco, are mainly in size, the males of the latter growing much larger, and also in the development of the skull and teeth. The male Hair Seal, when adult, weighs three or four times as much as the female, and is provided with enormous canine teeth, with which they fight terrible battles at the season of rutting, often injuring each other severely ; they are of a savage and dangerous disposition and are ugly antagonists even to man. On the expedition which captured those now in the Garden, one of the men was seized and so bitten and torn by a large bull Sea Lion that he died from the effects of his wounds.

They swim and float with great address, sleeping on the surface of the water ; they remain at sea during eight or nine months of the year, coming out on shore in vast numbers at the season of breeding, where they remain in some cases as much as three months without food or water. On land they progress with more ease than is common with other seals, by a gait somewhat like the canter of a horse ; they climb rocks easily and throw themselves from a height of ten or fifteen feet into the water or on the rocks without damage,—their tough skins and a layer of fat several inches thick, which lies immediately beneath, protecting them from injury. They are representatives of the family of Eared Seals (*Otaridæ*), all of which are of large size and are readily distinguished by the possession of an external ear, which is never more than an inch and a half long and is rolled tightly in the shape of a cone. There are seven or eight species of these seals, all being confined to the Pacific Ocean, where they range from the Arctic to the Antarctic regions, one species only being sometimes found up the Atlantic coast of South America as far as Brazil. The Fur Seals belong to this group; the undercoat of fur being very soft and thick in them and supplying the seal-skin of commerce.

The Seals now in the Garden were captured when quite young, in July, 1877, and were placed in the Garden in the following month ; they have not yet obtained their full size, but are growing rapidly and promise well for future development. When fully adult the males are seven or eight feet long and weigh from five to six hundred pounds ; the females weighing not over a fourth as much.

No. 10.—THE DEER HOUSE.

This building was completed in the spring of 1877, from the plans of George W. Hewitt, Esq., and affords accommodation at present for a somewhat varied assortment of herbivorus animals.

Here will be found the YAK (*Bison grunniens*); in a wild state native to the high mountain ranges and plateaus of Thibet and Tartary. There are several domestic breeds of the Yak used all over Central Asia for purposes of draught and burden. The long hair is used in the manufacture of various fabrics and the tails are much prized by the Tartars and Chinese, constituting among the former an insignia of rank when attached to the head of a lance ; by the Chinese they are dyed of various colors and used as fly-flappers.

The Wild Yak is a somewhat sullen and ill-tempered brute, and can use its long horns sideways with great effect.

THE ELAND (*Oreas canna*). This truly magnificent animal
is the largest of the Antelopes, the great home of which is in
Africa; this species being from the southern part. The Antelopes are generally of small or medium size, the Eland, however, is frequently of the size and weight of a large horse.
The venison is said to be of a delicious flavor and texture and
it is somewhat remarkable, considering the facility with which
the animal breeds under domestication, that they have not
been naturalized to a greater extent in Europe. It has been
found that they readily withstand the winters of France and
England, though the greater extreme of cold in our own
climate is more than they can bear without shelter.

THE OSTRICH (*Struthio camelus*)—during the warm weather
kept out of doors—will be found in winter occupying a large
pen on the east side of this building. This bird is the female
remaining of a fine pair which the Society purchased shortly
after the opening of the Garden.

They are natives of the hot, sandy plains in the interior of
Africa, over which they range in small flocks of rarely more
than half a dozen, subsisting mainly on a species of melon
which grows wild in those parts. The sexes are readily distinguished, the males having a mixture of black in their plumage, the females being of a grayish color. The Ostrich is
the largest known bird now existing, its head sometimes
reaching to a height of eight feet above the ground. Its
long legs give it great speed—some writers having estimated
its pace, when first startled, at fifty miles an hour. Its feet,
padded beneath, like those of the camel, adapt it to coursing
over the shifting, movable sands of its native regions without
sinking.

The wings are reduced to mere rudiments, as in all the
struthious birds, and are utterly useless for purposes of flight.
It is said, however, that the Ostrich raises them above the
sides and uses them as sails when—to use a nautical term—
running before the wind. The Ostrich is much prized by the
Bushmen, both for its eggs and feathers. Attempts at "Ostrich farming" are now being made in Lower California.

THE ORYX (*Oryx leucoryx*) is one of the innumerable
tribe of Antelopes inhabiting Africa. It is conspicuous for its
long, slightly curved, and tapering horns, which, as it is exceeding quick in its motions, it uses with much effect upon

an enemy. The lion has more than once been met and
pierced to the heart by these terrible horns when he thought
to secure, without danger, an unresisting prey. The species
is from the north-west of Africa.

THE SAMBUR DEER (*Cervus aristotelis*), from India,—a
fair type of the group of Rusine Deer, of which there are
several species, confined to Asia. The AXIS DEER (*Cervus
axis*), also from India, and the little SAVANNAH DEER (*Cer-
rus savannarum*), from South America, are also in this building during the winter.

KANGAROOS.

THE GREAT KANGAROO (*Macropus giganteus*).
THE RED KANGAROO (*Macropus rufus*).
THE DERBIAN WALLABY (*Halmaturus derbianus*).
BENNETT'S WALLABY (*Halmaturus bennetti*).

The Kangaroos inhabit the continent of Australia, as well as Van Diemen's Land and other of the adjacent islands; they, in common with nearly the whole fauna of the Australian region, belong to the order *Marsupialia*.

These animals derive their name from a pouch or bag situated on the lower part of the abdomen in the female, and containing the teats. The young animal being born—so to speak—prematurely, is in an undeveloped condition, and is at once placed by the mother in this pouch, where it attaches itself to a nipple and remains for some weeks, until it has attained a weight of several pounds, when it gradually begins to come forth. It does not permanently leave the bag until it has grown so large as to be of an inconvenient size for the mother to carry about. The order is a large one, containing nine-tenths of the fauna of the Australian region, and including the opossums of America. It presents a striking variety of habits and adaptations of form among its members; many of the other mammalian orders being represented by marsupial forms, which agree, more or less perfectly, with them in habits.

The Kangaroos fill a number of places in the economy of their native region—there being Brush Kangaroos, Rock Kangaroos, and Tree Kangaroos, all of which are equally at home in the surroundings indicated by their respective names.

The Wallabys are a sub-group of Kangaroos, differing slightly in structure.

THE RUFOUS RAT KANGAROO (*Hypsiprymnus rufescens*) is a small member of the family from New South Wales.

THE AOUDAD (*Ovis tragelaphus*), though classed among the sheep, differs considerably from the typical form of the family, and is often placed in a sub-group. It belongs in the north of Africa, where it ranges high up among the Atlas mountains, as our mountain sheep do among the Rocky mountains and the Sierras. The Aoudad is noticeable for its curving, powerful horns and for the thick beard which hangs from its neck to below the knee—almost to the ground.

THE AOUDAD.

The lower, or south-eastern portion of the Grounds is, as yet, unfinished, a considerable amount of grading and planting of trees being yet required to be done. The only enclosure which has been erected there is

No. 27.—THE POLAR-BEAR PEN,

Containing two fine young members of this species (*Ursus maritimus*). This animal is found throughout the Arctic regions of Europe, Asia, and America, rarely ranging below the fifty-fifth degree of latitude; how far to the north they find their way is unknown. Sir Edward Parry saw them at latitude eighty-two.

They measure sometimes nine feet in length from nose to tail and are dangerous visitors when pressed, as they often are, by hunger, to invade the camps of sailors, ice-bound in the northern seas. They live on fish, seals, and blubber, and being perfectly at home in the water, pursue their prey and capture it in its native element. The following account is given by an Arctic explorer of the cunning displayed by this animal in procuring food :—

"The Bear, on seeing his intended prey, gets quietly into the water and swims until to leeward of him, whence by short dives he silently makes his approach, and so arranges his distance that at the last dive he comes to the spot where the seal is lying. If the poor animal attempts to escape by rolling into the water, he falls into the bear's clutches ; if, on the contrary, he lies still, his destroyer makes a powerful spring, kills him on the ice, and devours him at leisure."

In captivity these animals live mostly on bread. The pair in the Garden were brought from Hamburg in December, 1876.

No. 11.—THE LAKE.

THE Lake, used in winter for skating, is occupied in summer by a number of aquatic birds, mostly belonging to the order *Anseres*. The following are usually to be seen :—

THE MUTE SWAN (*Cygnus olor*), Europe.

THE BLACK SWAN (*Cygnus atratus*), Australia.

THE MAGPIE GOOSE (*Anseranas melanoleucus*), Australia.

THE WHITE-FRONTED GOOSE (*Anser cærulescens*), North America.

THE CHINESE GOOSE (*Anser cygnoides*), China.

THE BRANT GOOSE (*Bernicla brenta*), Europe and North America.

THE CANADA GOOSE (*Bernicla canadensis*), North America.

THE DUSKY OR BLACK DUCK (*Anas obscura*), North America.

THE PINTAIL DUCK (*Dafila acuta*), Europe and North America.

THE AMERICAN WIDGEON (*Mareca americana*), North America.

THE SUMMER OR WOOD DUCK (*Aix sponsa*), North America.

THE CANVAS-BACK DUCK (*Fuligula vallisneria*), North America.

THE RED-HEADED DUCK (*Fuligula ferina var. americana*), North America.

THE EIDER DUCK (*Somateria molissima*), North Atlantic.

The last species, which yields much of the famous Eider down, is found along the Arctic coast of Europe and America. On the North Atlantic coast of this continent it sometimes winters as far south as New England. The down is plucked from the breast of the living bird and is very valuable. It is obtained, also, by robbing the nests, which the female parent lines with down, which, with her bill, she pulls from her own breast to make a soft resting-place for her offspring.

The Summer Duck differs from all the other true ducks of this country in its habit of living in trees,—its nest being commonly made in a hollow limb at a considerable distance from the ground. A group known as Tree Ducks, approaching somewhat to the geese, are found from Mexico to South America. The WHITE-FACED TREE DUCK (*Dendrocygna viduata*) and the RED-BILLED TREE DUCK (*Dendrocygna autumnalis*) are members of this group.

THE ADJUTANT (*Leptoptilus argala*), of India, is one of the extensive family of Storks, which are found throughout the world, with the exception of North America; it will be easily recognized by its long, thick bill. In its native country it wages successful war on the many venomous reptiles which there find a home, and also performs a part similar to that which is effectively taken in our Southern States by the Black Vulture. This is well stated in Jerdon's "Birds of India:"— "In Calcutta and some other large towns, the Adjutant is a familiar bird, unscared by the near approach of man or dog, and protected, in some cases, by law. It is an efficient scavenger, attending the neighborhood of slaughter-houses, and especially the burning grounds of the Hindoos, when the often half-burnt carcasses are thrown into the river. It also diligently looks over the heaps of refuse and offal thrown out into the streets to await the arrival of the scavenger's cart, where it may be seen in company with dogs, kites, and crows. It likes to vary its food, however, and may often be seen searching ditches, pools of water and tanks, for frogs and fish. In the Deccan it soars to an immense height in the air along with vultures, ready to descend on any carcass that may be discovered."

No. 26.—THE RABBIT WARREN.

PASSING around the Lake, the visitor reaches the Rabbit Warren, where are kept a variety of wild and domestic races of Rabbits,—among the wild species are generally the JACK-ASS RABBIT (*Lepus callotis*), of the plains of the Western States; the common GRAY RABBIT of our country (*Lepus sylvaticus*), and the COMMON HARE (*Lepus europæus*) of Europe.

One compartment of the Warren contains a number of WOODCHUCKS (*Arctomys monax*). This rodent, closely allied to the marmot of Europe, is well known to every farmer's boy from Canada to South Carolina; it burrows in the ground and when afforded proper facilities, as in this case, is rarely seen during the daytime. It is one of the most common of our wild animals, and may, perhaps, be better known under the name of Ground Hog—though why "Hog" it is difficult to say, as it does not resemble that animal in any way whatever.

No. 13.—THE MUSIC STAND.

IT is the general custom to have music on several afternoons in the week, during warm weather, in the Music Stand, opposite the Restaurant.

THE COMMON AMERICAN LYNX or WILD CAT (*Lynx rufus*) is distributed generally over the United States, and varies in color to so great an extent that different naturalists have insisted upon three or four not very well defined species within the range of its distribution. Though in appearance it is ferocious to the last degree, it is in reality a cowardly beast, and subsists altogether on small mammals and birds.

In the North it is replaced by the CANADA LYNX (*Lynx canadensis*), of somewhat larger size and grayer color. This species may also be known by the long tufts or pencils of hair which stand erect from the tip of the ear, and by its larger feet. As this Lynx is found far to the North among the snows of British America, it is undoubtedly true that nature—fitting all things for their necessities—has developed the enormous paw which is characteristic of the species, to answer the part of a snow-shoe in enabling the animal to range at will and capture its prey on the surface of the snow, without sinking enough to be impeded in its progress.

Like all the animals inhabiting the extreme North, which depend on an external covering for warmth, the fur of the Canada Lynx is exceedingly long and thick.

No. 14.—THE DEER PARK.

THE VIRGINIA DEER (*Cervus virginianus*) is the common Deer of the United States, and is found generally from the Eastern to the Gulf States, and from the Atlantic coast to the Missouri river. They are easily tamed, and breed readily under domestication.

The Llamas, which are kept in this Park, belong to the *Camelidæ*, and to a certain extent fill in South America the place which is held by the camels in Asia and Africa. They have long been domesticated, as Pizarro, on his conquest of Peru, found them in as general use as they are at the present time.

There are several wild species,—the others, so far as is known, being descendants of the wild stock.

THE LLAMA (*Lama peruana*) is much used by the natives of Peru to transport burdens up the steep passes of the Andes, and is one of the domestic races.

Among the wild ones are the HUANACO (*Lama huanacos*), about the size of the Llama, but of a reddish-brown color, and the VICUNA (*Lama vicugna*), a smaller variety, covered with reddish-brown wool. These, like the Alpaca, are much hunted

THE LLAMA.

for the valuable wool which they yield. They are domesticated with ease, and the Llama has bred several times in the
Garden.

THE MOOSE (*Alce americanus*), closely allied to the elk of
Northern Europe, is the largest of the Deer family, much exceeding in height the largest horse. This magnificent animal
formerly ranged into the upper parts of the Eastern and Middle States, but its numbers have lessened rapidly, and at present it is rarely found below the northern part of Maine, from
whence they range into all parts of British America.

The antlers of the Moose, at their fullest development, are
very widely palmated or flattened, and spread as much as five
feet from tip to tip. As with all of the Deer family, these
enormous horns are shed every year, early in the spring, and
are very shortly reproduced by a bony deposit from the blood.

It is almost beyond belief that so great a mass, weighing from
forty to sixty pounds, can be produced by such a process within
the short space of from ten to twelve weeks. During the season of rutting the bull Moose is a savage and dangerous animal, and it is well to keep beyond the reach of the terrible
blows which he deals with his sharp forehoofs. The animal
is, however, susceptible of a considerable degree of domestication, the writer having recently seen one which had been
broken to harness and trotted on the track. The gait of the
animal is a long, swinging trot, and is very rapid. The Society, at present, possesses a pair of full-grown Moose and two
young males.

THE WOODLAND CARIBOU.

Numerous attempts have been made in the Garden to keep
specimens of the WOODLAND CARIBOU (*Rangifer caribou*), but
in all cases the unsuitable climate and the impossibility of

providing the proper food have so far proved speedily fatal. The animal has an extremely northern range. There are two species, the one referred to reaching from Maine and New Brunswick westward to Lake Superior, and the BARREN LAND CARIBOU (*Rangifer grænlandicus*), far to the north in Greenland and Arctic America. They subsist for the most part on lichens, mosses, and small shoots and twigs of trees.

This is the only member of the Deer family in which the female as well as the male has antlers. The antlers are very irregular in development, and differ much in shape ; the tip and also the brow antler are generally palmated to some extent.

The Caribou represents in the New World the reindeer of the Old, and by training might be made useful to the Esquimaux as the latter is among the Lapps.

THE MULE DEER (*Cervus macrotis*) and the WHITE-TAILED DEER (*Cervus leucurus*) are both found on the plains of the United States, west of the Missouri river. The latter much resembles the common Deer, of which it is probably but a variety, while the former is considerably larger, and differs in the shape of its horns.

THE AUSTRALIAN CRANE.

In the Creek back of the Deer Park are usually a number of birds, some of which are placed in different buildings during the winter.

THE AUSTRALIAN CRANE, or NATIVE COMPANION (*Grus australasiana*), Australia.

THE SANDHILL CRANE (*Grus canadensis*), North America.

THE WOOD IBIS (*Tantalus loculator*), Southern United States.

THE WHITE IBIS (*Ibis alba*), Gulf States.

THE COMMON BITTERN (*Botaurus minor*), North America.

THE NIGHT HERON (*Nyctiardea grisea*), United States.

THE GREAT WHITE EGRET (*Ardea egretta*), Southern United States.

THE GREAT BLUE HERON (*Ardea herodias*), North America.

THE LEAST BITTERN (*Ardetta exilis*), United States.

These all belong to the order *Grallatores*, or Wading Birds. In them the legs are usually of great length, and are commonly bare of feathers for some distance above the tarsal joint; the neck is, in most species, of length proportioned to the legs. The order is a very extensive one, containing numerous forms distributed all over the world.

THE FULMAR PETREL (*Fulmaris glacialis*), the LAUGHING GULL (*Larus atricilla*), and the GREAT BLACK-BACKED GULL (*Larus marinus*), are members of the order *Gaviæ*, and are found along the coast of the Atlantic States; the former has a very extensive range to the north.

THE ALLIGATOR.

The large ALLIGATOR (*Alligator mississippiensis*) is generally to be seen during warm weather in one of the compartments of this Creek, either sunning himself on the bank or submerged in the water, nothing being visible but the tips of the ridges over the eye and the protuberance around the nostrils on the extreme end of the snout. This species is found in the rivers and bayous of the Gulf States, and is allied to the crocodile and gavial of Egypt and Southern Asia, and the caiman and jacare of South America. One species of CROCODILE (*Crocodilus americanus*) is also found in South America and Cuba.

No. 16.—THE CAMEL, ELK, AND BUFFALO PENS.

THE BACTRIAN, or DOUBLE-HUMPED CAMEL (*Camelus bactrianus*), and the COMMON CAMEL, or DROMEDARY (*Camelus dromedarius*), are both originally natives of Asia, where they now exist only under the subjection of man. The Bactrian Camel comes from the high, cold plains of Tartary, and is a more compact and powerful beast than the Dromedary, which comes from the warmer climate of Arabia, and is lighter and more fleet of foot. They are much used in the sandy deserts of Arabia and Northern Africa—nature having specially fitted their feet, like those of the ostrich, to the loose soil on which they walk, and also having provided them with a means of traveling for several days without requiring a fresh supply of water, part of the walls of the stomach supporting a double tissue, filled with cells, which absorb from the stomach a quantity of water sufficient to remain for some days as a reservoir, from which the necessities of the animal may be supplied. A number of Camels were imported by the United States Government, some years ago, with the idea of making them available in military operations in the West; but owing to the stony nature of much of the soil, for which their feet are not adapted, the experiment was not altogether successful. Those in the Garden are descended from this imported stock. .

Fossil remains of members of the *Camelidæ* have been found in the United States, thus proving that they were once indigenous to the country.

ALLEN·LANE·SCOTT

THE AMERICAN ELK.

THE AMERICAN ELK, or WAPITI (*Cervus canadensis*), is about the largest of the typical Deer, Judge Caton describing one, which lived for some time in his park in Illinois, that stood sixteen hands high at the withers, and was estimated to weigh nine hundred pounds; the average weight, however, of a full-grown buck would probably not be over six hundred. The Wapiti ranged originally all over North America and a large part of Canada; forty years ago a few were found in the mountains of Western Virginia and the wildest parts of New York, but civilization has gradually driven it, like the buffalo and the Indian, to a few fastnesses in the far West, where they yet make a stand before the final extermination which seems to inevitably await them. At the present time they range in small herds from the upper waters of the Missouri through the Yellowstone country westward to the Rocky mountains; they are found in fewer numbers South-west, in Texas, and a few are still left in the more secluded parts of

Michigan and Minnesota. They are readily kept, living on almost any kind of vegetable food, and are hardy and little liable to disease. Save in exceptional cases, and during the season of rutting, they are tractable and easily managed. The large buck in the collection had his antlers in the velvet when he was shipped to the Garden ; at this time the horn is soft and easily damaged. One horn was, unfortunately, broken, and has been reproduced each year successively in its damaged condition, consisting of a single fork about two feet long, while the other antler reaches its full development.

THE BISON.

THE AMERICAN BISON or BUFFALO (*Bison americanus*) may be observed to very good advantage in the large pen adjoining the Elk.

Of the geographical distribution of this species, past and present, Prof. J. A. Allen treats as follows in a "History of the American Bison," published by the Department of the Interior in 1877:—

" The habitat of the Bison formerly extended from Great Slave Lake, on the north, in latitude about 62°, to the north-eastern provinces of Mexico, as far south as latitude 25°. Its range in British North America extended from the Rocky mountains on the west to the wooded highlands about six hundred miles west of Hudson's Bay, or about to a line running south-eastward from the Great Slave Lake to the Lake of the Woods. Its range in the United States formerly embraced a considerable area west of the Rocky mountains—its recent remains having been found in Oregon as far west as the Blue mountains, and further south it occupied the Great Slave Lake basin, extending westward even to the Sierra Nevada mountains, while less than fifty years since it existed over the head-waters of the Green and Grand rivers, and other sources of the Colorado. East of the Rocky mountains its range extended southward far beyond the Rio Grande, and eastward through the region drained by the Ohio and its tributaries. Its northern limit, east of the Mississippi, was the great lakes, along which it extended to near the eastern end of Lake Erie. It appears not to have occurred south of the Tennessee river, and only to a limited extent east of the Alleghenies, chiefly in the upper districts of North and South Carolina.

" Its present range embraces two distinct and comparatively small areas. The southern is chiefly limited to Western Kansas, a part of the Indian Territory, and North-western Texas— in all together embracing a region about equal in size to the present State of Kansas. The northern district extends from the sources of the principal southern tributaries of the Yellowstone northward into the British possessions, embracing an area not much greater than the present territory of Montana. Over these regions, however, it is rapidly disappearing, and at its present rate of decrease will certainly become wholly extinct within the next quarter of a century."

Over near the whole of this country the Bison formerly ranged in vast herds, and the destructive side of man's nature cannot be better shown than in the contemplation of the comparatively small area to which they are now restricted. Yet even here they must be almost countless in numbers to withstand even for a short time the prodigious slaughter which goes on year after year among them. It has been estimated, by careful and competent authorities, that from the year 1870 to 1875 they were killed at the rate of two and one-half millions a year.

———•———

No. 17.—THE FOX AND WOLF PENS.

THE RED FOX (*Vulpes fulvus*) ranges in large numbers from the Carolinas northward into Canada, and from the Atlantic coast to the Missouri river. It is subject to much variation—the CROSS FOX (*Vulpes fulvus var. decussata*), a beautiful animal marked with two black stripes crossing each other on the shoulders, found from Canada into New York, and occasionally Pennsylvania, and the SILVER FOX (*Vulpes fulvus var. argentatus*), found sparingly in the North-western States—both being permanent and well-marked varieties. The latter is known by the beautiful and expensive furs which are made from its skin.

THE GRAY FOX (*Vulpes virginianus*) has a complete range throughout the United States, increasing in numbers towards the south, where it gradually replaces the Red Fox, which it much resembles in mode of life.

THE KIT OR SWIFT FOX (*Vulpes velox*) is the smallest of American Foxes, and is confined to the plains of the West.

THE ARCTIC FOX (*Vulpes lagopus*) is common to the Polar regions of the North. Like many species which inhabit those regions where the ground is covered by snow for a large part of the year, the fur of the Arctic Fox changes from a lead-brown color to white at the approach of winter. This provision of nature causes it to be less conspicuous against the snow and ice which surround it, and greatly aids it in obtaining food, as well as in escaping the necessity of serving as food to swifter and more powerful animals.

THE GREAT GRAY WOLF (*Canis occidentalis var. griseo-alba*) is the largest of American Wolves, and formerly ranged over the whole United States and Canada. The settlement of the country has, however, driven them, with other noxious beasts, to the more secluded forests and plains, where they

ALLEN·LANE·SCOTT

THE PRAIRIE WOLF.

are beyond the reach of man. The common color of the species is grayish-white, but it varies all the way from pure white to deep black.

THE PRAIRIE WOLF OR CAYOTE (*Canis latrans*) is well known to all Western travelers. Beyond the Missouri river they range in packs of from five or six to twenty, from Mexico well up into British America. They are intermediate in size between the Fox and Gray Wolf, and live mostly on the carcasses which are found upon the plains.

THE COMMON WOLF (*Canis lupus*) of Europe, resembles the Gray Wolf. A specimen in the Garden, from Italy, is smaller in size, being not much larger than the Cayote.

South America possesses several species of small Wolves, very fox-like in some of their characters. By some naturalists they have been constituted a group intermediate between the two. AZARA'S FOX (*Canis azaræ*) belongs to this group.

THE DINGO OR WILD DOG (*Canis dingo*) of Australia, was formerly supposed to be an aboriginally wild stock, but they are now taken to be descended from imported progenitors, which ran wild and have increased with great rapidity. They are wild, cowardly brutes, susceptible of little domestication, and cause by their depredations much loss to the sheep-raisers of Australia.

The Dogs, Wolves, and Foxes, with the Jackals, constitute a family of Carnivora known as the *Canidæ*.

No. 18.—THE WINTER HOUSE,

For tropical plants, is used merely to keep during cold weather those plants for which our winters are too severe for outdoor exposure. Being only a sort of storage-house, it is not open to visitors.

No. 19.—THE CATTLE PENS.

OPPOSITE the Wolves is an iron enclosure divided into pens in which are generally kept various members of the Ox and Deer families.

THE DOMESTIC GOAT (*Capra hircus*) is represented by many different breeds in all parts of the world. The Society recently imported several specimens of the celebrated CASHMERE GOAT. These animals are natives of Thibet and the adjacent countries, and are bred for the long, silky hair which covers them, and from which the famous Cashmere shawls and scarfs are made. They have been domesticated with some success both in Europe and this country.

THE ZEBU (*Bos indicus*). A number of breeds of these
cattle exist throughout China, India, and North Africa,
varying in size from that of a calf to the full adult of our
ordinary domestic cattle. They differ much in appearance,
there being breeds without horns, and some almost without
the characteristic hump on the shoulders, while in others the
horns are of great size, and some in which the hump weighs
from forty to fifty pounds.

The life of the Zebu is held sacred among the Hindoos,
and it is not uncommon for a particularly fine bull to be con-
secrated to the worship of Siva, and then turned loose to do
as he pleases among the natives, whose gardens he destroys
and whose homes he invades with perfect impunity.

They are much used as beasts of burden, and are also sad-
dled and ridden. They can be acclimated in this country
with a little care and breed readily. The Society now possesses a bull and four cows of a small variety, and a magnificent bull of the large, lop-eared breed, jet black in color,
contrary to the rule of his race, which are generally of a
mouse-gray.

THE PRONG-HORNED ANTELOPE (*Antilocapra americana*) is
remarkable on account of the formation of its horns in a
manner peculiar to itself alone. The horns of this species
resemble in appearance those of the hollow-horned ruminants,
in which the external covering of horny material grows around
a solid, bony core. These horns are never shed, and are not
replaced if lost by accident. In this Antelope, however, the
outside horny part is shed annually, and replaced, as in the
Deer; but with the important difference, that in the Deer
the antler is formed directly by a deposit from the blood,
while in the Antelope in question it is produced by growth
and hardening of the epidermis or outer layer of the skin.
The species is now confined to the plains of the temperate
regions of the West, where they are very common. They
are easily tamed, but are very delicate, and will not live for
any length of time under restraint of any character.

THE DOMESTIC SHEEP (*Ovis aries*), like the Goat in its
domesticated forms, is an inhabitant of the whole world.

In one of these enclosures are several YEMEN SHEEP, from
Persia, imported and presented to the Society by George

William Bond, Esq., of Boston. They are believed, by a cross with the common sheep of Spain, to have produced the celebrated Merino breed. These are, so far as is known, the first of the breed which have been brought to this country. They are white, with black heads.

THE FALLOW DEER (*Dama vulgaris*) is the common Deer of Europe. Its normal color is reddish-brown, spotted, like the Axis, with white. It is liable to variation in color, however, those in the Garden being pure white.

————+•————

No. 20.—THE REPTILE HOUSE.

THE accommodations in this building are inadequate to the proper display of the Society's rapidly increasing collection in this branch, and many of the cases are somewhat crowded.

The class *Reptilia* is composed of animals provided with lungs, a very simple digestive apparatus, and cold blood. Many of them live in the water, but are compelled to rise to the surface for the purpose of respiration. With the exception of a few of the Serpents, they are oviparous, and deposit their eggs on land. Reptiles proper are Turtles, Serpents, Lizards, and Saurians.

The members of the class *Batrachia* agree in many respects with the above. Their spawn, or eggs, is, however, hatched under water, and they are covered with a smooth skin in place of the scales with which reptiles are provided. The young live entirely in the water, and breathe with gills. In some genera, as *Siren* and *Menobranchus*, these gills are retained through life, and project from the sides of the neck, where they may readily be observed, the blood which fills them giving them a rich crimson color. In others, as the Frogs, Toads, Newts, and Salamanders, a complete metamorphosis takes place, the gills of the immature animal disappearing altogether in the adult. In the genus *Menopoma* the gills become reduced to a small orifice in the side of the neck, and the lungs are well developed.

Of the Batrachians, the SIREN (*Siren lacertina*), and the PROTEUS (*Menobranchus maculatus*) are usually in the collection. The former is found in the soft mud of streams and ditches from Georgia southward. The specimen in the Garden rarely comes out of the mud, except to get the worms on which it principally lives. The Menobranchus or Proteus is from the fresh-water streams and lakes of the Middle States, and feeds upon worms, grubs, and larvæ.

THE HELL BENDER or MUD DEVIL (*Menopoma alleghaniensis*) is found in the river mud of all tributaries of the Mississippi, and occasionally in other localities in the Gulf States.

The Salamanders and Tritons, or Newts, are found plentifully all over the United States. Many of them live entirely in the water, except at the breeding season. The RED-BACKED SALAMANDER (*Plethodon cinereus var. erythronotus*), the TWO-LINED SALAMANDER (*Spelerpes bi-lineatus*), the RED SALAMANDER (*Spelerpes ruber*), and the BLACK NEWT (*Desmognathus nigra*), are among the most common species.

The Toads and Frogs have a range co-extensive with the last. The COMMON Toad (*Bufo lentiginosus*) being found all over North America, with a number of well-marked varieties. The best known of the Frogs are the BULL FROG (*Rana catesbiana*), the SWAMP FROG (*Rana palustris*), the BROOK FROG (*Rana clamitans*), and the SHAD FROG (*Rana halecina*).

The North American fauna possesses a large and varied number of serpents (*Ophidia*), members of which order are found all over the world. The only venomous snakes belonging to North America are the different varieties of the Rattlesnake, the Moccasin, the Copperhead, and the Harlequin Snake. With the exception of the latter, these belong to the family *Crotalidæ*, distinguished by the presence of a deep pit between the eye and the mouth, and by the possession of a pair of poison fangs in the upper jaw, which are erectible at will.

The Rattlesnake is common to the whole United States, very rarely reaching into Canada, and becoming more plentiful towards the South, where they grow to a large size, sometimes reaching as much as six feet in length. They are characterized by a horny outgrowth of the epidermis at the end of the tail, known as the rattle, with which they make a whirring noise when excited, with the effect, it is presumed, of giving notice to their enemies that they are not to be trifled with without danger. The number of buttons, as the sections into which the rattle is divided are termed, has been commonly supposed to be a means of determining the age of the snake, but as they are frequently lost by accident, and are as frequently produced three or four at a time, it is evident that this belief, with a number of others equally well founded, which invest the popular mind regarding the serpent, may as well be abandoned.

THE BANDED RATTLESNAKE (*Crotalus horridus*) and the DIAMOND RATTLESNAKE (*Crotalus adamanteus*) are the most well-marked species of this genus. The former being most common in the Eastern and Middle States, while the latter ranges from North Carolina to Florida.

THE GROUND RATTLESNAKE (*Crotalophorus miliarius*) is a small species from the Southern States.

THE WATER MOCCASIN (*Ancistrodon piscivorus*) is confined to the wet and swampy lands throughout the South. A well-marked variety (*pugnax*) is confined to Texas.

THE COPPERHEAD (*Ancistrodon contortrix*) is found almost all over the United States, east of the Mississippi.

THE HARLEQUIN SNAKE (*Elaps fulvius*), of the Southern States, is also venomous, but in a lesser degree. It is of a very mild disposition, and has hardly ever been known to bite. It is one of the most beautiful of the order, being ringed with red, black, and yellow. The family, *Elapidæ*, to which it belongs, has the centre of its distribution in the Tropical Zone, throughout the whole circle of the earth, and includes some of the most deadly forms known.

THE RAINBOW SNAKE (*Abastor erythrogrammus*), Southern States.

THE KING SNAKE (*Ophibolus getulus*), Atlantic coast.

THE GREEN SNAKE (*Cyclophis vernalis*), Eastern and Southern States.

THE PINE SNAKE (*Pituophis melanoleucus*), south of New Jersey and Ohio.

THE CHICKEN SNAKE (*Coluber quadrivittatus*), Southern States.

THE CORN SNAKE (*Coluber guttatus*), Southern States.

THE MOUNTAIN BLACK SNAKE (*Coluber obsoletus*), United States.

THE BLACK SNAKE (*Bascanion constrictor*), United States.

THE WHIP SNAKE (*Bascanion flagelliformis*), Southern States.

THE RIBBON SNAKE (*Eutænia saurita*), Eastern and Southern States.

THE GARTER SNAKE (*Eutænia sirtalis*), North America.

THE WATER SNAKE (*Tropidonotus fasciatus*), Southern States.

THE WATER SNAKE (*Tropidonotus sipedon*), Eastern and Southern States.

THE HOG-NOSED SNAKE or SPREADING ADDER (*Heterodon platyrhinos*), United States, east of Mississippi.

These serpents all belong to what may be termed the Colubrine group. They are perfectly harmless to man, living on small birds, quadrupeds, worms, and insects; several genera —*Ophibolus* and *Abastor*—eating small snakes of their own and other species.

The Boas are a group of serpents inhabiting the tropical zone and attaining the largest size of any known members of the order, as the Anaconda and Boa, of South America, and the Python and Rock Snake, of Africa and Asia. They are possessed of great power and kill their prey by compression; they swallow without difficulty animals which appear larger in circumference than themselves, the articulation of their jaws and ribs permitting of a great degree of distension. Dr. Hartwig, in "The Tropical World," treats in a most entertaining style of these serpents:—" The kingly Jiboya (*Boa constrictor*) inhabits the dry and sandy localities of the Brazilian forests, where he generally conceals himself in crevices and hollows, in parts but little frequented by man, and sometimes attains a length of thirty feet. To catch his prey, he ascends the trees and lurks hidden in the foliage for the unfortunate agutis, pacas, and capybaras whom their unfortunate star may lead within his reach. When full-grown he seizes the passing deer; but in spite of his large size he is but little feared by the natives, as a single blow of a cudgel suffices to destroy him. Prince Maximilian of Neu Wied tells us that the experienced hunter laughs when asked if the Jiboya attacks and devours man. The Sucuriaba, Anaconda, or Water Boa (*Eunectes murinus*), as it is variously named, attains still larger dimensions than the constrictor, as some have been found of a length of forty feet. It inhabits the large rivers, lakes, and marshy grounds of tropical America and passes most of its time in the water, now reposing on a sand bank with only its head above the surface of the stream, now rapidly swimming like an eel, or abandoning itself to the current of the river. Such is its voracity that Firmin ('Histoire Naturelle de Surinam') found in the stomach of an Anaconda a large sloth, an iguana nearly four feet long, and a tolerably sized ant-bear, all three nearly in the same state as when they were first swallowed—a proof that their capture had taken place within a short time.''

Several specimens will be found in the collection, of the COMMON BOA (*Boa constrictor*) of South America, and the TREE BOA (*Epicrates angulifer*) of Cuba. The genus Epicrates belongs to a section of the family which are distinctively known as Tree Boas ; they are rarely more than seven or eight feet long, are arboreal in habit, and are found in the West Indies and Guiana.

A large variety of Lizards are distributed throughout the world, being most common in the warmer parts of the temperate zone. There are many species, belonging to the order *Lacertilia*. Among the most common North American forms are :—

THE STRIPED LIZARD (*Eumeces fasciatus*), Middle Atlantic and Southern States.

THE SIX-LINED LIZARD (*Cnemidophorus sex-lineatus*), Southern States.

THE BROWN LIZARD (*Sceloporus undulatus*), Southern States.

THE HORNED TOAD (*Phrynosoma cornuta*), South western United States and Mexico.

THE CHAMELEON (*Anolis principalis*), Southern States.

These are all small and harmless, many of them living among trees and feeding upon worms, insects, &c.

The Iguanas also belong to this order ; in some species they attain a length of four and five feet ; pass most of their time in trees and live on fruit and birds' eggs. They inhabit tropical America and the West Indies, where they are considered as very good eating by the natives.

The Society possesses a . very good collection of typical forms of North American Turtles. These consist of the Turtles proper, which are generally of large size and live in the sea ; the Terrapins, which live in streams, ponds, and marshes ; and the Tortoises or Land Turtles. There are a number of forms belonging exclusively to the American fauna, of which the following are on exhibition :—

THE LEATHERBACK TURTLE (*Thassalochelys caouana*), Atlantic coast.

THE GREEN TURTLE (*Chelonia mydas*), Atlantic coast.

THE SOFT-SHELLED TURTLE (*Aspidonectes ferox*), Gulf States.

THE SNAPPER TERRAPIN (*Chelydra serpentina*), North and South America.

THE MUSK TERRAPIN (*Aromochelys odoratus*), Eastern and Southern States.

THE MUD TERRAPIN (*Cinosternum pennsylvanicum*), Eastern and Southern States.

THE RED-BELLIED TERRAPIN (*Pseudemys rugosa*), Middle Atlantic States.

THE FLORIDA TERRAPIN (*Pseudemys concinna*), Southern States.

THE SALT WATER TERRAPIN (*Malacoclemmys palustris*), Atlantic and Gulf States.

THE CHECKERED TERRAPIN (*Chrysemys picta*), Eastern and Southern States.

THE CHICKEN TERRAPIN (*Chrysemys reticulata*), Gulf States.

THE SPECKLED TERRAPIN (*Chelopus muhlenbergii*), Pennsylvania and New York.

BLANDING'S TORTOISE (*Emys meleagris*), Western United States.

THE BOX TORTOISE (*Cistudo clausa*), Eastern and Southern States.

THE GOPHER TORTOISE (*Testudo carolina*), Southern States.

THE GREAT OR ELEPHANT-FOOTED TORTOISE (*Testudo elephantopus*), of the Galapagos Islands, is represented by a pair which, with the other Tortoises, are in a compartment of the Rabbit Warren.

THE INDIAN FRUIT BATS (*Pteropus medius*), known also by the name of Rousette Bat and Flying Fox, are temporarily kept in this building, although they belong to the mammalian order *Cheiroptera*. They exist in large numbers in India and the neighboring islands, where they grow to a very large size, the expanded wings sometimes measuring four or five feet from tip to tip. Sir Emerson Tennent gives the following account of some of their habits :—

"A favorite resort of these bats is to the lofty India-rubber trees, which on one side overhang the Botanic Garden of Paradenia, in the vicinity of Kandy. Thither for some years past they have congregated, chiefly in the autumn, taking their departure when the figs of the *ficus elastica* are consumed. Here they hang in such prodigious numbers that frequently large branches give way beneath their accumulated weight.

"Every forenoon, generally between the hours of 9 and 11 A. M., they take to wing, apparently for exercise, and possibly to sun their wings and fur and dry them after the dews of early morning. On these occasions their numbers are quite surprising, flying as thick as bees or midges.

"After these recreations they hurry back to their favorite trees, chattering and screaming like monkeys, and always wrangling and contending angrily for the most shady and comfortable places in which to hang for the rest of the day protected from the sun.

"The branches they resort to soon become almost divested of leaves, these being stripped off by the action of the bats attaching and detaching themselves by means of their hooked feet. At sunset they fly off to their feeding-grounds, probably at a considerable distance, as it requires a large area to furnish sufficient food for such multitudes."—*Natural History of Ceylon.*

When at rest, the Fruit Bat hangs head downward, by one foot, wrapping itself tightly in the folds of its wings.

Another mammal, also kept in this building during cold weather, is the SIX-BANDED ARMADILLO (*Dasypus sex-cinctus*). The Armadillos belong to the order *Edentata*—so called from the imperfections in their supply of teeth. They live in the warmer portions of the New World, from Texas into South America. They burrow in the ground and live on worms and insects.

The order includes, among existing animals, the Armadillos, Sloths, and Ant-eaters of tropical America and Africa. Some of the largest of extinct mammals, of which remains have been discovered, as *Glyptodon, Mylodon,* and *Megatherium* were also Edentates,—the first having been a sort of gigantic Armadillo fifteen feet long.

No. 21.—THE AVIARY.

PENDING the construction of a new Aviary at the southeastern end of the Garden, between the Lake and Thirty-fifth street drive, the collection of birds will be found much crowded.

The order *Passeres,* or Perching Birds, is the largest and most comprehensive of all the higher groups and includes almost all our songsters. In fact, a very large proportion of the smaller birds which are familiar in our midst will be found classed within its limits.

The Thrushes are represented here by the WOOD THRUSH (*Turdus mustelinus*) of America, the ROBIN (*Turdus migratorius*), the ENGLISH BLACK BIRD (*Turdus merula*), the MOCKING BIRD (*Turdus polyglottus*), the CAT BIRD (*Turdus carolinensis*), and a number of other species.

THE REED BIRD, RICE BIRD, or BOBOLINK (*Dolichonyx oryzivorus*) is well known to epicures. They migrate in vast numbers from South to North at the approach of summer and back again towards autumn, at which time they become very fat on the ripened seeds of the reeds which grow on marsh lands along the rivers near the coast, and are shot in great numbers as a table delicacy.

·THE LONG-TAILED WEAVER BIRD (*Chera progne*). This species may be known by the great elongation of the centre tail feathers of the male. These reach so great a length that a celebrated African traveler says of them :—" I am informed that in the breeding season, when the male has assumed his nuptial livery and long tail feathers, his flight is so labored that the children constantly run them down. They are quite unable to fly against the wind and in rainy weather can hardly be got to move out of the thick bushes in which, knowing their helplessness, they conceal themselves.

"The Kaffir children stretch bird-limed lines across the fields of millet and Kaffir corn and snare great numbers of the males by their tails becoming entangled in the lines."—— *Layard, "Birds of South Africa."*

The Weaver or Whidah Birds are noted for the peculiar nests which they weave from grass; these are mostly built on the community system, hundreds of the birds constructing together a sort of roof under which they separately build their nests. These nests are of different shapes, some of them much resembling a chemist's retort, with the neck down, the orifice serving as entrance for all the birds whose dwellings are within. They are all natives of Africa.

THE COMMON MYNAH (*Acridotheres tristis*), and the BROWN MYNAH (*Acridotheres fuscus*) belong to an East Indian group allied to the starlings.

THE MAGPIE (*Pica caudata*), well known by its thievish propensities, is found all over Europe; it belongs to the *Corvidæ* or Crows, as do the WHITE-BACKED PIPING CROW (*Gymnorhina leuconota*) and the BUTCHER CROW (*Barita destructor*), both of Australia, and the BLUE JAY (*Cyanurus cristatus*) of North America.

The order *Picariæ* is represented by the Woodpeckers, Kingfishers, Cuckoos, Toucans, and Trogons.

THE LAUGHING JACKASS.

THE LAUGHING JACKASS or GIANT KINGFISHER (*Dacelo gigantea*) is the largest of the Kingfishers, and inhabits Australia. It differs somewhat in habits from most of the immediate group to which it belongs, living in the woods, frequently far from water; its diet is also more mixed than is customary with its kind, as it eats not only fish, but small quadrupeds, birds, and reptiles.

Its common name is derived from its cry, which has a striking resemblance to a hoarse laugh.

The common Kingfisher of our country, which is often seen sitting motionless on a branch over the water, watching intently for a small fish to pass within its reach, is the BELTED KINGFISHER (*Ceryle alcyon*).

Quite a number of Cuckoos are in existence throughout the world, very few being natives of America. THE CHAPPARAL COCK (*Geococcyx californianus*), also known as ROAD RUNNER, from the extraordinary speed with which it runs, and the YELLOW-BILLED CUCKOO (*Coccyzus americanus*), belong to this family.

Among the Trogons are numbered some of the most beautiful and gorgeous birds of the American fauna. THE CUBAN TROGON (*Prionotelus temmnurus*), a richly and highly colored member of the family, will be found in the collection ;—it is a very rare bird, the difficulty of keeping them alive making them to be one of the most unusual birds met with in a collection. Ramon de la Sagra thus describes its habits in his " Histoire de l'Isle de Cuba " : —

".This Trogon, one of the most splendid members of its family, has as yet been met with only in the Island of Cuba, of which it is not the least beautiful ornament. Very common in the woods, its favorite place of abode, its plaintive song may be heard there in the evening, but especially in the morning, repeated at lengthened intervals. The first portion of the note is highest and longest, and may most readily be imitated with a trumpet. It is this habit that has induced the Guaranis of Paraguay to say, speaking of another species, that it cries in the morning that the sun may rise, and in the evening because it is setting. Dwelling alone in the vast woods, it perches generally on the lower branches of the trees, and there remains immovable for hours at a time, apparently asleep, or, at least, insensible to what is going on around it. It is, therefore, easily shot and many are killed for the table, its flesh being very good." There are some fifty species of Trogons, found in the warm regions of America, India, and Africa.

THE TOCO TOUCAN (*Ramphastos toco*).

CUVIER'S TOUCAN (*Ramphastos cuvieri*).

The Toucans are found only in tropical America, and will be readily recognized by the size and brilliant coloring of their bills,—large and unwieldy as these seem to be, they are in reality very light, being entirely filled with a honey-comb of air cells. The plumage is richly colored and has a peculiar satin-like softness of texture. The Toucans are in a measure carnivorous, and often prey upon smaller birds.

" Common as these birds are in their native wilds, it is exceedingly difficult to detect their breeding places ; it is certain that they deposit their eggs in the hollow limbs and holes of the colossal trees so abundant in the tropical forests, but I was never so fortunate as to discover them. * * * * In their manners, the *Ramphastidæ* offer some resemblance to the *Corvidæ* and especially to the Magpies ; like them they are very troublesome to the birds of prey, particularly to the owls, which they surround and annoy by making a great noise, all the while jerking their tails upwards and downwards. The flight of these birds is easy and graceful, and they sweep with ease over the loftiest trees of their native forest." (Prince Maximilian of Wied.) There are a number of species, beautifully illustrated by Mr. Gould in a " Monograph of the Ramphastidæ."

The *Psittaci*, comprising the Macaws, Parrots, Parrakeets, and Cockatoos, is a large and varied order of birds, found throughout tropical America, Asia, Africa, and Australia. Among them are some of the most splendid specimens of their class ; many of them learn to talk and imitate various sounds with great facility, and they are much kept as pets.

The Macaws are exclusively American in distribution and have been referred to at page 36.

The most beautiful birds, perhaps, of the order, come from the Australian region, where are found a great number of species. Among them are PENNANT'S PARRAKEET (*Platy-cercus pennantii*), the ROSEHILL PARRAKEET (*Platycercus eximius*), the PALE-HEADED PARRAKEET (*Platycercus pallidiceps*), BAUER'S PARRAKEET (*Platycercus zonarius*), SWAINSON'S LORIKEET (*Tricoglossus novæ-hollandiæ*), and the BLOOD-RUMPED PARRAKEET (*Psephotus hæmatonotus*). The ZEBRA or GRASS PARRAKEET (*Melopsittacus undulatus*), also from Australia, is one of the most beautiful of these birds,—it is very small, and being of a green color, marked with undulating yellowish-white lines, bordered with black, it is almost impossible to distinguish it from the grass in which it is generally to be found. These little birds differ from all the other Parrots, in the faculty which they alone possess, of uttering a repetition of several notes which might almost be termed a song.

THE RING-NECKED PARRAKEET (*Palæornis torquatus*) and the ALEXANDRINE PARRAKEET (*Palæornis alexandri*) are very handsome and graceful birds from India.

The Parrots of the New World, as a rule, are not so brilliantly colored as those species from the other hemisphere, being generally green, with various markings of red, yellow, and white.

Those species which are usually in the collection are the YELLOW-FRONTED AMAZON (*Chrysotis ochrocephala*), the YELLOW-CHEEKED AMAZON (*Chrysotis autumnalis*), the GOLDEN-NAPED AMAZON (*Chrysotis auripalliata*), the WHITE-FRONTED AMAZON (*Chrysotis albifrons*), and the WHITE-EARED PARROT (*Conurus leucotis*). North America possesses one species, the CAROLINA PARROT (*Conurus carolinensis*), which formerly ranged up to North Carolina and Kentucky, but which is now rarely found even in the Gulf States.

A number of small Parrakeets are also common in South America, as the CAYENNE PARRAKEET (*Brotogerys tuipara*), the YELLOW-THROATED PARRAKEET (*Brotogerys tovi*), and the PASSERINE PARRAKEET (*Psittacula passerina*), the latter much resembling the Love Birds of Africa.

The Cockatoos are confined to the Australian region. Being for the most part large birds of graceful flight, their appearance is large flocks is described as being singularly

beautiful. The following will be found in this building:—
The SULPHUR-CRESTED COCKATOO (*Cacatua galerita*), the
ROSEATE COCKATOO (*Cacatua roseicapilla*), LEADBEATER'S
COCKATOO (*Cacatua leadbeaterii*), and the ROSE-CRESTED
COCKATOO (*Cacatua moluccensis*). These birds are said to
live to a great age,—a specimen of the last-named species,
on a perch on the east side of the building, being supposed
to be over eighty years old.

THE CRESTED GRASS PARRAKEET (*Calopsitta novæ-hollandiæ*)
is a very beautiful bird from Australia, living in hollow
trees. Unlike most of its order, it breeds with some facility
in confinement, when afforded proper accommodations.

The order *Columbæ*, comprising the Pigeons and Doves,
and which also included the now extinct DODO (*Didus in-
eptus*), is a very extensive group, containing some three hundred species, more than one-third of which are natives of
the Malayan Archipelago, the remainder being distributed
over the world. A large number of species are indigenous
to North America, many of which have been exhibited at
the Garden ; of these, the following are usually on hand:—

THE WHITE-CROWNED PIGEON (*Columba leucocephala*),
Southern United States and West Indies.

THE CUBAN PIGEON (*Columba inornata*), West Indies.

THE COMMON WILD PIGEON (*Ectopistes migratorius*), North
America.

THE BLUE-HEADED PIGEON (*Starnænas cyanocephala*),
Southern United States and West Indies.

THE CAROLINA DOVE (*Zenædura carolinensis*), North
America.

THE ZENAIDA DOVE (*Zenaida amabilis*), West Indies.

THE KEY WEST DOVE (*Geotrygon martinica*), Southern
United States and West Indies.

These birds are all more or less common in the regions
where they are found. The Common Wild Pigeon is mi-
gratory in its habit, and travels in enormous flocks of many
millions. They may be seen in some parts of the West, during their annual migration, covering acres of trees at night,
when roosting.

Many of the Pigeons from the Malayan Islands, where is the great home of the order, are of striking beauty, among them are the BLOOD-BREASTED PIGEON (*Phlogœnas cruentata*), of the Phillipine Islands; the BRONZE-WINGED PIGEON (*Phaps chalcoptera*), of Australia, and the GOURA or CROWNED PIGEON (*Goura coronata*), of New Guinea.

THE CROWNED PIGEON.

This superb bird is as large as a Guinea fowl, and has the top of its head surmounted by a beautiful crest of radiating feathers. It readily bears the winters of France and England, and has frequently laid eggs, which in many cases have been hatched, in both of those countries.

THE BARBARY TURTLE DOVE (*Turtur risorius*); the HALF-COLLARED DOVE (*Turtur semi-torquatus*), of Africa; the BARRED DOVE (*Geopelia striata*); and the GROUND DOVE, of the Southern United States and the West Indies (*Chamœpeleia passerina*), belong also to this group.

The order *Gallinæ* includes the Guinea, Turkey, Curassow, Guan, Pheasant, and Partridge, and contains a large proportion of those birds which are known as "game birds," alike esteemed by the sportsman and the gourmand. They are mainly terrestrial birds. Some of them roost in trees, but during the day-time live on the ground.

The Curassows and Guans belong exclusively to America, ranging from Mexico across the Isthmus of Panama down to the southern part of Brazil. They are large birds, living on fruits and seeds, and are said to be very well adapted for the table. They do not, however, breed with any facility in this latitude. Among them are—

THE CRESTED CURASSOW (*Crax alector*), Guiana.

DAUBENTON'S CURASSOW (*Crax daubentoni*), Central America.

THE GLOBOSE CURASSOW (*Crax globicera*), Central America.

THE GALEATED CURASSOW (*Pauxi galeata*), South America.

GREEY'S GUAN (*Penelope greeyi*), New Granada.

THE WHITE-FRONTED GUAN (*Penelope jacucaca*), Brazil.

The WILD TURKEY (*Meleagris gallopavo*) of North America; the GUINEA FOWL (*Numidia meleagris*), originally from Africa, but now domesticated all over the world; the COMMON PEA FOWL (*Pavo cristata*), native to India, but, like the Guinea, naturalized everywhere; and the JAVAN PEA FOWL (*Pavo muticus*), from Java and the Malay Peninsula, distinguished from the foregoing by the peculiar shape of the plumes on the head, and by the rich green color on the breast—all belong to the group of Pheasants (*Phasianidæ*).

Sir Emerson Tennent states that the Common Pea Fowl abounds to such an extent in the Island of Ceylon that their harsh cries at early morning are a great source of annoyance to the inhabitants. He also adds that the bird, as known in its domesticated state in other countries, gives but a very inadequate idea of its size and magnificence when seen in its native forests.

The true Pheasants are indigenous to Asia and its islands; but the matchless beauty of their plumage, and the delicate quality of their flesh, has caused some of them to be largely introduced into Europe. The bird now known as the ENGLISH PHEASANT (*Phasianus colchicus*) originally came from Western Asia, and is believed to have been introduced into England about the close of the tenth century. It has long been perfectly naturalized, and is one of the most beautiful of

the family. Many of the Pheasants have those feathers which
lie immediately around the base of the tail—known as the
upper tail coverts—immensely elongated, forming a large fan,
like the train of the peacock, which they have the power to
erect at will, thus exposing a surface of brilliant and beautiful
coloring, which in many species is without parallel in the
animal kingdom. These plumes are developed to this extent
only in the males, and from the fact of their being displayed
frequently in the presence of the female during the breeding
season, it is supposed that they play an important part in
attracting her at this period.

The Silver Pheasant (*Euplocamus nycthemerus*) and the
Golden Pheasant (*Thaumalea picta*) are natives of China,
but have been largely acclimated elsewhere.

The Grouse are an allied group, generally distributed
throughout the northern hemisphere. A number of them
are natives of America, where they are commonly, though
erroneously, called *Pheasant*.

The Ruffed Grouse (*Bonasa umbellus*) and the Pinna-
ted Grouse or Prairie Chicken (*Cupidonia cupido*) are two
of the best known species.

There is much confusion still existing among naturalists as
to the relationships which should properly be recognized between the Partridges found on the opposite sides of the
Atlantic,—some holding them to be distinct sub-groups,
while others claim that the differences are not susceptible of
definition to the extent of warranting a separation.

Our well-known Common Partridge, or Quail, as it is
frequently called (*Ortyx virginianus*), is the most widely-
distributed species in North America, and has several marked
varieties in the South-west and in Cuba. The female is lighter
in color than the male, and has the buff of the neck replaced
by white.

On the Pacific coast several genera have the head beautifully ornamented with plumes of feathers rising in various
shapes. Examples of this form are the Plumed or Mountain
Partridge (*Oreortyx pictus*), and the members of the genus
Lophortyx from Arizona and California.

The domestic fowl is the most widely distributed member of the *Gallinæ*, being spread in various breeds over the whole world. The progenitor of this invaluable bird is supposed to be the wild BANKIVA COCK (*Gallus bankiva*), which is a native of India.

There are several species of Tinamous in South America, one of which, the CINEREUS TINAMOU (*Tinamus cinereus*) may be seen in the collection. These birds present some peculiarities of structure so strongly marked that they have been placed in an order, *Crypturi*, by themselves. Mr. Darwin, speaking of the Tinamou in "The Zoology of the Voyage of the Beagle," states that it approximates somewhat to the habits of the grouse, but that it rarely rises from the ground, and may be readily caught by a stick having a noose at the end.

THE COMMON TRUMPETER (*Psophia crepitans*), Guiana.

THE CAYENNE RAIL (*Aramides cayennensis*), South America and West Indies.

THE CAROLINA RAIL (*Porzana carolina*), North America.

THE FLORIDA GALLINULE (*Gallinula galeata*), Gulf States.

THE MARTINIQUE WATER HEN (*Porphyrio martinica*), Southern United States and West Indies.

THE BLACK-BACKED PORPHYRIO (*Porphyrio melanotus*), Australia.

THE COMMON COOT (*Fulica americana*), North America, and the AMERICAN WOODCOCK (*Philohela minor*), Eastern United States, all belong to the order *Grallatores* or Wading Birds. They live along the borders of water-courses and streams, from which they pick out the small fishes, insects, and worms which serve them as food.

No. 22.—THE PRAIRIE-DOG VILLAGE.

THE PRAIRIE DOG (*Cynomes ludovicianus*) is a small, burrowing rodent, much resembling the Spermophiles, which are common throughout the western part of the United States.

THE PRAIRIE-DOG VILLAGE.

They are found in enormous numbers over the expanse of open country between the Missouri river and the Rocky mountains—the villages formed by them sometimes occupying miles of country, which is completely honey-combed with their burrows ; they dig to a considerable depth, those in the Garden having burrowed under a wall fourteen feet deep ; they were overcrowded, however, and in a state of nature, with room to stretch out their villages on every side, it is not probable that they dig so deeply. The Dogs are found in close association with the burrowing owl and the rattlesnake, which has given rise to the supposition that they all live together on terms of intimacy and friendship. This is far from being the case, however, the evidence going to prove that the snake invades the home of the Dog for the purpose of feeding upon the young, while the owl—to save itself the trouble of digging its own habitation—takes possession of the deserted burrows which are left in the gradual change of location which is continually going on among the Dogs. Strong evidence was given of a natural enmity existing between the two, by the introduction of a pair of the owls into the enclosure of the Dogs at the Garden,—they were instantly attacked by the latter, and as their wings had been clipped, they were unable to get away, and although they fought desperately were finally killed.

A small colony of the BURROWING OWLS (*Spheotyto cunicul-aria var. hypogæa*) is on the side of the walk directly opposite the Prairie Dogs. They are found on the plains west of the Mississippi river and also in South America,—this bird being a variety of the South American form.

Next to the Owls is a cage containing some STRIPED GOPH-ERS (*Spermophilus tridecem-lineatus*). The Spermophiles form a large sub-group of the squirrel family ; they live in burrows in the ground, but are directly connected with the tree squirrels by almost insensible gradations of form, one of which, the LINE-TAILED or MEXICAN ROCK SQUIRREL (*Spermophilus gram-murus*), will be seen in a cage in the Carnivora House. It is difficult to distinguish this species, without close examination, from the true squirrels, which live in trees. The Striped Gopher is found from Western Michigan to the Missouri river and south to Arkansas.

RICHARDSON'S SPERMOPHILE (*Spermophilus richardsonii*) and the GRAY GOPHER (*Spermophilus franklinii*) are also North American forms.

THE COMMON SKUNK (*Mephitis mephitica*) is a small car-nivore belonging to the Musteline group; it inhabits the United States from the Missouri river eastward,—the western and south-western parts of the country being infested by several other species, one of which ranges into South America. The true Skunks are confined to the New World ; in them the anal glands characteristic of all the *Mustelidæ* reach their most complete development, and secrete a fluid which the animal can eject at will to a distance of several feet, and which possesses an odor at once the most powerful and intol-erable of any known animal secretion. The fur of the Skunk is very long and fine, and is much worn under the euphoni-ous designation of "Alaska Sable."

THE BADGER (*Taxidea americana*) belongs to the same group as the Skunk ; it lives in burrows, which it excavates with its powerful claws, and is found through the western United States up to about latitude fifty-eight degrees in Brit-ish America, though it is seldom seen east of the Missouri river ; southward it is replaced by a well-marked variety (*ber-landieri*). Allied species of Badger are found in Europe and Asia.

Among the true Squirrels (*Sciuridæ*) of North America, those generally to be found in the collection are the SOUTH-ERN FOX SQUIRREL (*Sciurus vulpinus*), of the Gulf States; the CAT SQUIRREL (*Sciurus cinereus*), from New Jersey to Virginia; the GRAY SQUIRREL (*Sciurus carolinensis*), United States east of the Missouri river, and the RED SQUIRREL (*Sciurus hudsonius*), which ranges from British America to the Gulf States.

The Gray Squirrel is subject to much variation, the Black as well as the White Squirrel belonging to this species.

————•————

No. 23.—THE PHOTOGRAPH STAND

Will be passed on the right of the walk on the way out, close by the Monkey House. Photographs of many of the animals in the collection can be purchased here at very moderate prices.

————•————

No. 24.—THE SUN-DIAL

Points out the time of day at Philadelphia and at numerous other points on the earth's surface.

ALPHABETICAL LIST OF ANIMALS IN THE GARDEN.

(80)

www.ingramcontent.com/pod-product-compliance
Lightning Source LLC
Chambersburg PA
CBHW021952190326
41519CB00009B/1229